The Grand Unified Theory of Physics

By

Joseph M. Brown

Basic Research Press

The Grand Unified Theory of Physics

By
Joseph M. Brown
First Edition
First Impression

ISBN: 0-9712944-6-1
Published By
Basic Research Press
120 East Main Street
Starkville, MS 39759
United States of America

The Grand Unified Theory of Physics

Contents

The Grand Unified Theory of Physics

Preface

This book presents a whole new paradigm for physics. It presents a unified mechanism for deriving all the primary observables in physics. It presents a mechanical model of the neutrino, it shows a mechanism for the fine structure constant and shows why it pervades all of physics, it shows how fundamental particles have a constant value of angular momentum, and it shows the structure of a proton, how its mass, angular momentum, strong nuclear, weak nuclear, and charge fields are produced. A structure of the electron is developed which shows how its mass is held together, how it produces the charge field, and how it produces angular momentum. The book presents the structure of the neutron which gives evidence of how the weak nuclear force functions, and shows the special relativity mechanisms for mass-energy equivalence, mass growth with velocity, matter shortening with velocity, and time dilation. It shows why the mechanism of mass growth of matter with velocity gives matter waves and shows that the waves produce magnetism by the same mechanism that the proton and other charged particles produce their electrostatic field. The book shows that atoms and neutrons produce gravitational fields by a mechanism similar to the breathing sphere model which produces electrostatic fields. The amplitude of the breathing sphere is controlled by, and is equal to, the basic ether particle radius. Further, the same mechanism controlling the breathing sphere amplitude is believed to remove one basic ether particle from a photon for each wave of travel that it executes, which gives the illusion of an expanding universe.

The fine structure constant, 1/137.036, is the ratio of the electromagnetic force to the nuclear force. It also is the velocity of the lowest energy electron orbit in a hydrogen atom in velocity of light units. It pervades all of quantum electrodynamics. However, the number has been a mystery since it was discovered more than seventy years ago. In this model for a grand unified theory of the universe, everything is made up of kinetic particles. The gas of these particles has a root mean square speed that is eight percent larger than the mean speed. We show a model in which the electromagnetic speed (the speed of light) is the difference of these speeds. Also, the same model gives the strong nuclear speed as the background mean speed. Forces generated in a kinetic particle universe, of course, are a function of the square of the speed. The square of the ratio of these two speeds, the root mean square speed less the mean speed to the mean speed is 1/137.109 and thus clearly must be the ratio of the electromagnetic force to the strong nuclear force.

Einstein's special theory of relativity uses a space-time continuum and predicts that as velocity increases, the mass of matter will increase, the length of matter will shorten, and time for processes will increase. Further, the energy content of matter is its mass times the square of the speed of light. The Einstein system is almost universally accepted in science. Many physicists believe it is impossible to derive the theory of relativity observations from classical Newtonian mechanics. In this paper we present a system of absolute space with a separate absolute time, a purely classical (Newtonian) system, from which the above four phenomena are derived. The system used is a kinetic particle system. The model immediately gives the equivalent energy of mass. The model also gives the wave properties of matter in motion.

Magnetism is known to be due to charges in motion. We present a kinetic particle mechanism which produces the electrostatic force, produces the deBroglie wave property of matter, and shows that the deBroglie wave generates the same mechanism which produces the electrostatic force to produce the magnetic force.

The model of the proton structure and its formation, which we present, leads into a hypothesized structure for the electron. With this structure of the electron the structure of the neutron is indicated. We thus present structures for the most basic assemblage of particles, which is the neutrino, and we derive structures for the proton, electron, and neutron.

The mechanism producing gravity is similar to that producing electromagnetism. When the electron is formed the portion of its structure producing the electrostatic field matches the proton electrostatic field except always in the opposite directions. The result is that the flows are matched except for the diameter of a basic ether particle. So, the two fields move with a half amplitude equal the ether particle radius with respect to each other. This then produces the gravitational field. Quantum electrodynamic effects are the result of each elementary matter charge particle consisting of a discrete mass which orbits at the speed of light and produces waves in the background. Matter in motion has a wave path as a result of being accelerated by an eccentric mass. A photon is a narrow "string" of mass which extends completely over one wave length. The particles in the waves have velocities with magnitude and direction. A wave function ψ is a complex number and can be used to describe the expected velocity (magnitude and direction) of a matter or radiation particle. The function ψ is called the probability amplitude. When an event occurs with two possible paths, 1 and 2 then $|\psi_1 + \psi_2|^2$ gives the probability of the combined event. This is the basis of the derivation of the Schroedinger equation. We illustrate the use of probability amplitudes in the analysis of the partial reflection of light and in diffraction gratings.

All of these results taken together can only lead to the conclusion that the universe is made up of Newtonian kinetic particles.

Joseph M. Brown

120 East Main Street

Starkville, MS 39759

United States of America

August, 2004

The Grand Unified Theory of Physics

1. The Postulates

Are there fundamental postulates from which all of physics can be derived? If so, what are these postulates? We have discovered that everything in the universe can be constructed of one type small elastic particle. Yes, just one type particle makes all other particles of matter, makes light and all other radiation, and even makes the elusive neutrinos. We call the basic particle the brutino. The name brutino means tiny brute, since it is very small and is the brute that makes everything. The postulates are:

1. Space is three dimensional.
2. One type of particle makes everything
3. The particle is smooth, elastic, moves, and collides with other particles.

Our first significant discovery toward developing this theory was the discovery of the mechanism of the mysterious fine structure constant.

2. The Fine Structure Constant

Over 30 years ago Brown, Harmon, and Wood [1][1] published a paper entitled "A Note on the Fine Structure Constant." At that time we were employed by the McDonnell Douglas Aircraft Company (now the Boeing Company) and were doing research to develop an advanced propulsion system. The effort centered on developing a theory of gravity based upon a postulated system consisting of an inert hard particle gas ether. The particles are very small with a diameter $\left(\doteq 10^{-35}\,m\right)$ many billions times smaller than a proton, with a mass $\left(\doteq 10^{-66}\,kg\right)$ many billions times less than an electron, a quantity of particles on the order of 10^{85} per cubic meter, and a mean free path of the ether gas[2] on the order of $10^{-16}\,m$. Stable assemblages of these particles were presumed to be neutrinos and these, in turn, formed the observed particles and radiation in the universe. Chapter 5 and Appendix B discuss the structure of the neutrino.

The results of this research (over the years 1967-9) were published by Brown and Harmon [3] in 1972. One of the most interesting results of this research was the observation that the ether Maxwell-Boltzmann distributed gas mean speed v_m and root mean square speed v_r arranged in the form $\left[(v_r - v_m)/v_m\right]^2$, gives the numerical result $(v_r/v_m - 1)^2$ = $\left(\sqrt{3\pi/8} - 1\right)^2$ = 0.0072934814 = 1/137.108733, see page 21 of [3]. We noticed the closeness of this to the fine structure constant α=0.007297352533(27) = $1/137.03599976(50)$, [4]. We thus suspected that the square root of this number might be the velocity coupling ratio for the

---

[1] The references are listed at the end of the book.

[2] The values of the ether parameters were developed in Brown [2], but are revised and presented in Appendix A.

electromagnetic field. This number times the speed of light thus should be the orbital velocity predicted for the electron in the lowest energy orbit of the hydrogen atom.

The orbital velocity is computed now. Let m_e and m_p be the relativistic masses of the electron and proton. The force balance gives $e^2/\left(r_e+r_p\right)^2 = m_e \mathrm{v}_e^2 / r_e$. Where e is the charge of the electron, r_e is the distance from the electron to the center of mass, and r_p is the distance from the proton to the center of mass. Further $\hbar = m_e r_e \mathrm{v}_e + m_p r_p \mathrm{v}_p$, where $\hbar$ is Planck's constant and v_p is the proton orbital velocity. Combining the above equations and solving for v_e/c gives[1]

$$\mathrm{v}_e / c = \frac{e^2}{\left[\hbar c\left(1+m_e/m_p\right)\right]} \tag{2.1}$$

The relativistic mass ratio is given by

$$\frac{m_e}{m_p} = \frac{m_{eo}}{\sqrt{1-\left(\mathrm{v}_e/c\right)^2}} \frac{\sqrt{1-\left(\mathrm{v}_p/c\right)^2}}{m_{po}} = \frac{m_{eo}}{m_{po}} \frac{\sqrt{1-\left(\mathrm{v}_e/c\right)^2\left(m_e/m_p\right)^2}}{\sqrt{1-\left(\mathrm{v}_e/c\right)^2}} \tag{2.2}$$

Where m_{eo} and m_{po} are the electron and proton rest masses. Note that m_e/m_p appears on both sides of the equation and that v_e/c appears twice in the equation.

The above equation for m_e/m_p and the previous equation for v_e/c can be solved by eliminating m_e/m_p. The resulting equation is a quartic algebraic equation. Alternatively it is noted that the term $\sqrt{1-\left(\mathrm{v}_e/c\right)^2\left(m_e/m_p\right)^2}$ in (2.2) can be taken as unity and v_e/c in this

---

[1] This refinement was first noted by Dr. Darell B. Harmon, Jr.

same expression can be taken as 0.007293 and the result will be sufficiently accurate. Thus from the above equation we obtain

$$\frac{m_e}{m_p} = \frac{m_{eo}}{m_{po}} \frac{1}{\sqrt{1-(0.007293)^2}} = 1.0000266 \frac{m_{eo}}{m_{po}}. \tag{2.3}$$

Thus, the relativity correction is an increase in mass of 27 parts per million.

To account for relativity in (2.1) the value m_e / m_p is the non-relativistic value 0.00054461702, from reference [4], times 1.0000266. Thus the term $1 + m_e / m_p$ is

$$1 + m_e / m_p = 1 + 1.0000266(0.00054461702) = 1.00054463. \tag{2.4}$$

The value of v_e / c now is

$$v_e / c = \frac{e^2 / (\hbar c)}{1.00054463}. \tag{2.5}$$

Substituting $e^2 / (\hbar c)$ as α gives

$$v_e / c = \frac{\alpha}{1.00054463} = 0.0072933803 \tag{2.6}$$

Comparing this with $\left[(v_r - v_m) / v_m \right]^2$ we have 0.0072934814 vs. 0.0072933803 or, the orbital velocity (in speed of light units) obtained from the Maxwell-Boltzmann ether theory is 1 part in 70,000, or 14ppm, larger than obtained from the fine structure constant. The difference is not due to experimental error but possibly may be due to the physical interpretation of the measurements. Nonetheless, the numerical agreement of the quantities is interesting.

Possible reasons for the theoretical and experimental difference could be that the center of charge and the center of mass of the electron could differ enough to produce the error and the non-circularity of the electron path could produce the effect.[1] If it is assumed that $\left(\mathrm{v}_r / \mathrm{v}_m - 1\right)$ is the electromagnetic velocity coupling ratio then its square is (0.0072934814 ÷ 0.0072933803 = 1.00001386 or) 14ppm larger than the "measured" value. The theoretical velocity ratio $\left[\left(\mathrm{v}_r\text{-}\mathrm{v}_m\right)/\mathrm{v}_m\right]$ presumes that the electron path is a perfect circle while actually the electron could undulate as a result of its wave nature (for waves of higher frequency than the deBroglie wave) and there could be an effect due to the elliptic path of motion. If the experimental velocity were increased to account for these effects then it is possible that the theoretical and experimental difference would be less.

What we have from the expression $\left(\mathrm{v}_r - \mathrm{v}_m\right)$ divided by v_m is a numerator equal to a force speed less a transport speed, and a denominator equal to a transport speed. Research throughout the following 30 years (1970-2000) evolved kinetic particle mechanisms for the term $\left(\mathrm{v}_r - \mathrm{v}_m\right)$ and for the denominator term v_m, [2].

The fine structure constant often is described as the ratio of the electromagnetic force to the nuclear force. Recognizing that forces, in the assumed kinetic particle universe, are direct functions of velocities squared the implication here is that the "electromagnetic force velocity" would be the speed of light "c" and thus $\mathrm{v}_r - \mathrm{v}_m = c$ and the electromagnetic force would be proportional to $\left(\mathrm{v}_r - \mathrm{v}_m\right)^2$. The strong nuclear force velocity then would be v_m and the strong nuclear force would be proportional to $\mathrm{v}_m{}^2$. We now describe kinetic particle mechanisms to produce both the "electromagnetic velocity" and the "strong nuclear velocity."

---

[1] The effect of magnetism in hydrogen between the two moving charged particles compared to the electrostatic force is $(\mathrm{v}_p/c)(\mathrm{v}_e/c) = (\alpha \mathrm{v}_e/c)(r_p/r_e)(\mathrm{v}_e/c)$ $= \alpha^2\, m_e/m_p = 2.90\times10^{-8}$ times the electrostatic charge force.

In order to have observable phenomena in a hard particle ether universe it is necessary to have a basic organizing mechanism. As a prelude to describing an organizing mechanism let us first describe a related *disorganizing* phenomenon.

Consider a perfectly elastic cubic box containing a gas of N identical elastic spheres of mass "m" uniformly distributed throughout the box. All have exactly the same speed "v". Further, initially $1/6N$ of these particles move to the right, $1/6N$ move to the left, $1/6N$ move upward, $1/6N$ move downward, $1/6N$ move inward, and $1/6N$ move outward. The initial mean speed, of course, is "v". The initial root mean square speed also is "v", since all speeds are the same. Without any interference the particles will collide and take a Maxwell-Boltzmann speed distribution. The initial energy is $(1/2)\mathrm{N}m\mathrm{v}^2$ and the final energy is $(1/2)\mathrm{N}m\mathrm{v}^2$. Thus $\mathrm{v}_r = \mathrm{v}_r' = \mathrm{v}$, where the prime indicates the final condition. The initial linear momentum is zero. However, in a Maxwell-Boltzmann distribution the mean speed is given in terms of the root mean square speed by $\mathrm{v}_m = \sqrt{8/(3\pi)}\,\mathrm{v}_r = 0.92\mathrm{v}_r$. The mean speed dropped by 8 percent in this process.

We will now describe a possible organizing mechanism, again see Chapter 5 and Appendix B. Consider a tornado-like flow of the gaseous particles which has rotation coupled with a sink-source flow. The rotation of some assemblages would be right-handed and the rest would be left-handed. Assume that the inflow to the sink is a gas that becomes completely "condensed" so that the particles are touching each other to produce a solid moving core translating and rotating like a rifle bullet and the particle speeds, just before the final condensation, are an unchanged (speed) random sample of the Maxwell-Botlzmann distributed background, and that the inflow to this condensed "core" is axial along the doublet axis.[1] The particle inflow velocity will then be v_m, the background mean speed. The particle inflow energy will be proportional to v_r^2, the square of the background root mean square velocity. Consider an imaginary cylindrical tube defining the condensed core. If energy is conserved during this final condensation process the flow velocity must increase from v_m to v_r, the opposite of the disorganizing process described above. Thus, an axial thrusting force must be occurring on the side of the condensed core which is proportional to $(\mathrm{v}_r - \mathrm{v}_m)^2$.

This thrust propels the assemblage. The assemblage has inflow into the core at velocity v_m, an outflow at velocity v_r, and the outflow circulates via the doublet flow back to the inlet. The whole assemblage moves at the velocity $\mathrm{v}_r - \mathrm{v}_m$. All observables in the universe are assumed to be made up of these assemblages so all observed phenomena are due to assemblages moving at the speed of light. The assemblages presumably are formed with a range of masses, which mass is all in the core. The core is all contained within a sphere with a diameter of one mean free path. The angular momentum and thrust presumably are dependent only upon the

---

[1] The size of the condensed core is estimated by assuming essentially a radial constant density inflow to a sphere with a radius equal to the mean free path, l, at which radius the gas reaches the background mean speed v_m. Independent of the shape of stream tubes inside this mean free path radius if the flow is assumed spherically symmetric, there is a spherical space where the gas is completely condensed (i.e. all particles are touching others). The mass inflow rate $\dot{m}$ at the mean free path sphere is $\rho \mathrm{v} a = m_b \eta (\mathrm{v}_m) 4\pi l^2$. The mass inflow at the condensed sphere radius is $\rho \mathrm{v} a = \left[m_b / 4/3\,\pi r_b^2 \right] \mathrm{v}_\mathrm{m} 4\pi r_v^2$. In these expressions m_b is the basic particle mass $\left(10^{-66}\,kg\right)$, r_b is the particle radius $\left(10^{-35}\,m\right)$, η is the background particle number density $\left(10^{85}/m^3\right)$, and r_v is the core sphere radius. Solving this gives $r_v \doteq 2\times10^{-26}\,m$. This is a rough estimate of the minimum size core, its mass is

$$m_v = \left[\left(4/3\pi r_v^3 \right) / \left(4/3\pi r_b^3 \right) \right] m_b = \left(2\times10^{-26} / 10^{-35} \right)^3 \times 10^{-66} \doteq 10^{-38}\,kg.$$

free-field properties and therefore do not vary with the core mass. As the condensation begins to take place by background particles flowing into a sink the particles can almost reach sonic speed by flowing inward along straight stream tubes. After reaching this speed it is necessary for the stream tubes to curve. Curved stream tubes produce thermal velocity separation of the particles because of the centrifugal force on the particles. Such thermal separation occurs in Ranque-Hilsch vortex tubes, see Lay [5]. This thermal separation due to inertia may be a part of the mechanism which permits complete condensation of a gas. In any case, going down a curved path causes rotation and thus produces angular momentum. Such an assemblage as envisioned here produces an angular momentum depending only upon the mean free path and other background ether properties, thus causing all observable elements to have the one constant value of angular momentum.

In order for such an assemblage as envisioned to occur it almost certainly must be contained principally within a region extending a distance in the order of the mean free path "l". If the structure were much larger than one mean free path the assemblage would be disrupted by the outflow. Consistent with this requirement we assume that the inflow is radial down to the surface of a sphere with a radius equal the mean free path and, at that location, the flow velocity is v_m. Actually, the stream tubes will be curving by then, but for our analysis we will assume all the rotatory component has been achieved by an average radius of $0.8l$. Further, we assume the time the mass is inside the mean free path radius sphere is for a length of πl divided by v_m. Now, the angular momentum of the assemblage is the flowing mass inside the sphere, which is $\left(\rho_0 4\pi l^2 v_m\right)\left(\pi l/v_m\right)$ times its effective radius $0.8l$ times its velocity v_m. Thus

$$\frac{\hbar}{2}=\left(\rho_0 4\pi l^2 v_m\right)\left(\pi l/v_m\right)(0.8l)\,v_m \tag{2.7}$$

or

$$\hbar = 6.4\pi^2 \rho_0 v_m l^4 \tag{2.8}$$

The values of all these parameters are developed in Appendix A. Substituting the values in (2.8) gives

$$\hbar = 6.4\pi^2 \left(1.0495\times10^{19}\right)\left(3.5103\times10^{9}\right)\left(8.2045\times10^{-17}\right)^4 \tag{2.9}$$

$$=1.0544\times10^{-34}\, kg\, m^2/s$$

This agrees closely with the measured value, of course, since we used this equation to determine l, see Appendix A.

One specific value of core mass for an assemblage moving at the speed of light in a circular path can produce an angular momentum of $\hbar/2$ and have the centrifugal force balance the thrust. This particle would model a proton. The value of this thrust is quite large – and can easily be computed. The balance of centrifugal force with the thrust F is $F = m_p c^2/r_p$ and the proton angular momentum is $\hbar/2 = m_p c r_p$. Substituting r_p from the angular momentum equation into the force equation gives

$$F = 2m_p^{\,2} c^3/\hbar = 2(1.67262158\times10^{-27})^2(299792458)^3 \div 1.054571596\times10^{-34} \tag{2.10}$$

$$=1.42959\times10^{6}\, N$$

This is a very large force!!

The orbiting assemblage, with a tangential speed $v_r - v_m$ ($= c$, of course,) produces two fields. One field is due to the continuous inflow at velocity v_m into the core sink (which is moving in a circular path at velocity $v_r - v_m$). Another field is the similar outflow translating at velocity v_r from the core source (moving at $v_r - v_m$). The orbiting flow of the basic assemblage produces the magnetic moment.

A flow meter, in principle, could move around fixed to the orbiting assemblage at such a location that it always experienced an inflow at the maximum speed v_m. This field associated with this inflow is the strong nuclear force field. At a distance "r" of many assemblage circular path diameters a flow meter fixed with respect to the matter particle would experience principally a spherically symmetric flow field which is oscillating inward at velocity v_m and outward at velocity amplitude v_r, the source output velocity. Thus the field is an alternating flow of velocity amplitude $v_r - v_m$. This field is the electromagnetic field.

In order to understand the particle interactions resulting from the above described flow fields, let us review some experimental and theoretical work on particles immersed in a medium. The following results are taken from Whittaker [6].

In 1876, C.A. Bjerknes immersed two identical spheres in water, had them oscillate in a breathing mode, and measured the mutual force of interaction between the spheres. The force was found to be an inverse square force, as expected, and it was attraction when the oscillation was in-phase and repulsion when out-of-phase. A theoretical analysis was developed which explained the results. This analysis is presented in the book by Bassett [10] and is repeated in Appendix C. Further, the theoretical results were extended to gaseous media and to a large variety of oscillation producing devices – all with the inverse square force result.

Now, consider the long-range force between two charged nucleons (say a proton and proton, or proton and anti-proton). The basic assemblage (i.e., the neutrino) producing the proton has rotation of one handedness and the anti-proton has rotation with the opposite handedness. It is assumed that like rotations would result in out-of-phase waves and unlike rotations would result in in-phase waves, to give repulsion of like charges and attraction of unlike charges respectively. The field strength would be measured by an oscillating flow field of velocity $v_r - v_m$. This spherically symmetric oscillating flow field is the mechanism producing the electromagnetic force.

Two nucleons at short range could have their assemblage circular paths oriented and phased in such a way that they would experience only the inflow at velocity varying from 0 to v_m (and avoid the outflow velocity v_r). The nuclear force strength then is proportional to $v_m{}^2$. This is the strong nuclear force mechanism.

In the above envisioned universe the strongest possible attractive force is produced by a flow velocity v_m. The coupling velocity is v_m, the coupling force constant is unity for nuclear forces. For the electromagnetic force the coupling velocity is v_r - v_m, the coupling velocity constant is $(v_r - v_m)/v_m$, and the coupling force constant would be

$$[(v_r - v_m)/v_m]^2 = \left(\sqrt{3\pi/8} - 1\right)^2 = 0.0072934814 = 1/137.109 \qquad (2.11)$$

This is close in magnitude to the fine structure constant, α . In addition to the good quantitative agreement between $[(v_r - v_m)/v_m]^2$ and the fine structure constant there is another qualitative feature of this kinetic particle theory of the universe which lends credence to the validity of the theory. All observables in the kinetic particle theory must stem from the mechanism for producing the speed of light and the mechanism gives the speed of light as the rms speed less the mean speed of the all-pervading ether gas. This mechanism, of course, pervades the whole universe.

To better appreciate the significance of the fine structure constant we quote the following from Feynman [7].

> "Immediately you would like to know where this number [α] for a coupling comes from: is it related to pi, or perhaps to the base of natural logarithms? Nobody knows. It's one of the greatest *da--* mysteries in physics: a *magic* number that comes to us with no understanding by man.

You might say the "hand of God" wrote the number, and
'we don't know how He pushed His pencil'."

The significance of the fine structure constant also is discussed by Lévy-Leblond [8]. We believe the agreement of the Maxwell-Boltzmann derived number and the fine structure number can not be dismissed as chance. The kinetic particle theory of the universe must mirror reality.

3. Relativity and the Wave Property of Matter

In this book we have assumed a universe of identical elastic spherical particles which particles make up a gaseous ether and make up all matter and radiation. Further, we assume the gaseous ether is in a three dimensional space with a separate absolute time. The particles are very small with a diameter in the order of the Planck length $\left(\doteq 10^{-35} m\right)$ and with a mass billions and billions times smaller than the electron. The particle number density is very large $\left(\doteq 10^{85} m^{-3}\right)$ and the mean free path is on the order of nuclear particle diameters.

A photon is assumed to be a stable assemblage of a large number of these ether particles translating at the speed of light, of course. Each fundamental matter particle at rest such as a proton, is assumed to be a neutrino which is a very small stable assemblage (again made up of very many basic particles) moving at the speed of light in a circular path with a very small diameter.

From these above assumptions we immediately have the result that the energy of a matter particle at rest is the matter particle mass M_0 (which is the background particle mass times the number of particles making up the matter particle) times the square of the speed of light. Defining energy as mass times the square of the mass's velocity then

$$E_0 = M_0 c^2 \tag{3.1}$$

which is the famous Einstein formula for the "equivalence" of mass and energy.

In order to accelerate matter a series of photons bombard the matter and each photon is partly scattered and partly captured.[1] The captured parts of the photons are the mass that is added to the matter to

[1] This process involves absorbing the impacting photon and then emitting a lower energy photon.

accelerate it. When a force does work on a particle of mass m and accelerates m from zero velocity to v, the work done is $\frac{1}{2} m \mathrm{v}^2$ so that $\frac{1}{2} m \mathrm{v}^2$ is the particle's kinetic energy change. If a mass moving at velocity v (having energy $m\mathrm{v}^2$) is absorbed by another mass moving at the same magnitude and direction of velocity v then the energy change of the increased mass particle is $m\mathrm{v}^2$. Consider now a photon of energy $e_\gamma = m_\gamma c^2$. This can be written in terms of its linear momentum p_γ as $e_\gamma = (p_\gamma / c)c^2$. Consider the acceleration of a matter particle due to the result of scattering photons. Let M_o be the mass of the matter particle at rest and let M_v be the mass when moving at velocity v. The matter particle energy when moving is $M_\mathrm{v} c^2$ and the linear momentum is $P_m = M_\mathrm{v} \mathrm{v}$. The linear momentum P_γ of the photons (assuming many photons were used) is $P_\gamma = M_\gamma c$, where M_γ is the sum of the photon "masses". Let the scattered photon total "mass" be $M_s = kM_\gamma$ so that the captured "mass" is $M_c = (1-k)M_\gamma$. The momentum transferred by the captured mass of each photon is $m_\gamma(1-k)c$, where m_γ is the "mass" of one photon. The momentum transferred by the scattered portion of each photon is $m_\gamma kc$, since the scattering is spherically symmetric with the maximum back scatter momentum being $2m_\gamma kc$ and the minimum forward scatter is zero for an average of $m_\gamma kc$. Thus, the momentum transferred to the matter particle is all of each photon's momentum. The total momentum imparted then is the sum of the initial momentum of each photon. Let us denote the total momentum as *P*. The differential energy change for the matter particle is the force times the distance so

$$dE_\mathrm{v} = Fdx \tag{3.2}$$

where E_v is the energy of the moving matter particle and F is the force applied. The force is the time rate of momentum change of the matter particle which, also, is the time rate of momentum change of the group of impacting photons.

$$F = \frac{dP}{dt} \tag{3.3}$$

Now

$$dE_v = Fdx = (dP / dt)dx = vdP \tag{3.4}$$

We also have

$$dP = d(M_v v) = M_v dv + v dM_v \tag{3.5}$$

and

$$dE_v = vdP = v(M_v dv + v dM_v) \tag{3.6}$$

further

$$dE_v = d(M_v c^2) = c^2 dM_v \tag{3.7}$$

Thus

$$c^2 dM_v = v\left(M_v dv + v dM_v\right) \tag{3.8}$$

Simplifying

$$\frac{dM_v}{M_v} = \frac{vdv}{c^2 - v^2} \tag{3.9}$$

Integrating M_v from M_o to M_v and v from o to v gives

$$ln\frac{M_v}{M_o} = -\frac{1}{2}\ln\frac{c^2 - v^2}{c^2} = \frac{1}{2}\ln(1-\beta^2) \tag{3.10}$$

where $\beta = v / c$. Now

$$\frac{M_v}{M_o} = \frac{1}{\sqrt{1-\beta^2}} \tag{3.11}$$

We see that this is the well known mass growth equation and note it has been derived from classical Newtonian mechanics which uses an absolute space with a separate absolute time system.[1]

Let us determine the portion of the photon mass that is captured and that which is scattered. The captured mass is

$$M_c = M_v - M_0 = M_0\left(1-\sqrt{1-\beta^2}\right)\Big/\sqrt{1-\beta^2} \tag{3.12}$$

The momentum balance relates the scattered and captured mass by the equation[2]

$$(M_s + M_c)c = (M_c + M_0)\text{v} \tag{3.13}$$

Thus

$$M_s = (M_c + M_0)\beta - M_c = M_0\left[\beta - 1 + \sqrt{1-\beta^2}\right] \div \sqrt{1-\beta^2} \tag{3.14}$$

Now

$$M_s/M_c = \beta\Big/\left(1-\sqrt{1-\beta^2}\right)-1 \quad M_s/M_c = \beta\Big/\left(1-\sqrt{1-\beta^2}\right)-1 \tag{3.15}$$

Some values of M_s/M_c versus β are now obtained.

β	0.01	0.02	0.1	0.5	0.8	0.9	0.99
m_s/m_c	199	99.0	18.9	2.73	1.0	0.595	0.153

From this table we note that at small velocities practically all the mass is scattered, while at large velocities practically all the mass is captured.

The fact that the velocity of matter can never exceed the speed of light results simply from the fact that the accelerating agent (i.e. the photon) is moving at the speed of light.

When a photon interacts with matter at rest the circular path becomes a spiral path as seen from a rest frame. However, in a frame moving at the translational velocity v of the particle the spiral is seen as a closed path and

---

[1] This analysis was developed for this theory by Dr. Darell B. Harmon, Jr.
[2] This results since the average scattered mass has its velocity at 90° to the impacting velocity.

since angular momentum is constant along the path, the closed path is an ellipse. Figure 3.1 shows the elliptic path of a matter particle moving to the right at velocity v_0. The two foci F_1 and F_2 are shown. The mass "m" takes the path shown by the ellipse in the reference frame moving at velocity v_0.

Since the mass always moves at velocity c (the speed of light) the mass at point A moves at velocity $c - v_0$ in this frame and the mass at B moves at velocity $c + v_0$. Angular momentum conservation requires that the mass at A times $(c - v_0)$ times the distance A to F_2 be the same as the same mass at B times $(c + v_0)$ times the distance F_2 to B. The major semi-axis is "a" and the eccentricity is "*e*", as shown in Figure 3.1. Angular momentum conservation gives

$$(a + ae)(c - v_0) = (a - ae)(c + v_0) \tag{3.16}$$

Solving for *e* gives

$$e = v_0 / c = \beta \tag{3.17}$$

Thus, the eccentricity is the particle translational velocity in speed of light units.

We will now determine the relation between the orbit shape and the particle velocity. Figure 3.2 shows an elliptic orbit of a particle moving to the right at velocity v_0. An ellipse is the locus of points where the distance from the point of a fixed focus (i.e. distance $\overline{AF_1}$) added to the distance from that same point to the other focus (i.e. distance $\overline{AF_2}$) is constant. For example, if a string of length $\overline{F_1A}$ plus $\overline{AF_2}$ is fixed at F_1 and F_2 and a pencil is placed inside the string and a trace is made the trace will be an ellipse.

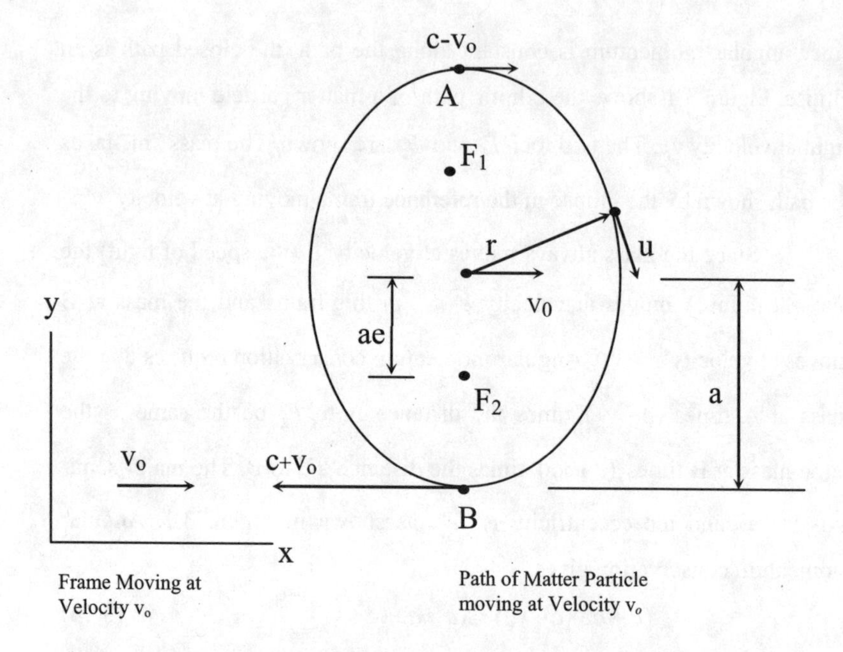

Figure 3.1. Matter Particle Moving at Velocity v_0

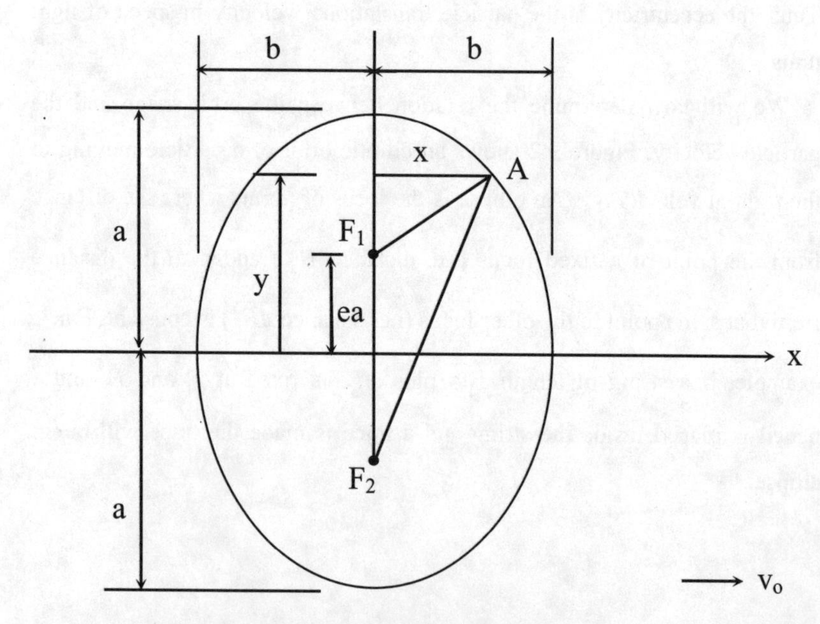

Figure 3.2. Elliptic Orbit Geometry

The length of the string is given by

$$\overline{F_2A}+\overline{AF_1}=\sqrt{(ea+y)^2+x^2}+\sqrt{(y-ea)^2+x^2}=2a \tag{3.18}$$

which simplifies to

$$(a^2-a^2e^2)(a^2-y^2)=a^2x^2 \tag{3.19}$$

or

$$(1-e^2)(a^2-y^2)=x^2 \tag{3.20}$$

when

$$\mathrm{y}=0 \quad \mathrm{x}=\pm\,\mathrm{b} \tag{3.21}$$

then

$$a^2(1-e^2)=b^2 \tag{3.22}$$

Thus

$$b/a=\sqrt{1-e^2} \tag{3.23}$$

Since $\mathrm{e}=\mathrm{v}_0/c=\beta$, from (3.23), we have

$$b/a=\sqrt{1-\beta^2} \tag{3.24}$$

This is the ratio of the minor axis to the major axis and clearly shows the orbit size reduction. Every matter particle in a piece of matter, such as a bar of steel, experiences this shortening with velocity. Thus, the complete bar will be shortened in the direction of motion by the factor $\sqrt{1-\beta^2}$. We therefore have

$$l_{\mathrm{v}}/l_o=\sqrt{1-\beta^2} \tag{3.25}$$

The velocity "u" of mass "m" on this elliptic path at radius r from F_2, as shown in Figure 3.1, is given in [9] by the equation

$$u^2=g\left[2/r-1/a\right] \tag{3.26}$$

where "g" is a constant (=GM by McClusky in [9]). The maximum velocity is when $r=a-ea=(1-e)a$ and has the value $c+\mathrm{v}_0$. From this

$$(c+v_0)^2 = g\left[\frac{2}{(1-e)a} - \frac{1}{a}\right] = \frac{g}{a}\frac{1+e}{1-e} \tag{3.27}$$

and

$$g = (c+v_o)^2 a(1-e)/(1+e) \tag{3.28}$$

Let the time for an orbit, i.e. the period, be τ_v (from [9]), substituting the value of GM as g from the above, and using e as β gives

$$\tau_v = 2\pi a^{3/2}\Big/\sqrt{g} = 2\pi a\Big/\left(c\sqrt{1-\beta^2}\right) \tag{3.29}$$

When $v_o = 0$ (i.e. when $\beta = 0$) the period $= 2\pi r/c$ -- obviously the circumference of the circle divided by the speed of light. The period increases with motion and grows without bound as $\beta\left(=v_0/c\right)$ approaches unity, or as the velocity approaches the speed of light. Nuclear particles, which disintegrate and emit radiation and produce other matter particles, are observed to decay slower when moving – and governed by the law, $\tau_v/\tau_0 = 1/\sqrt{1-\beta^2}$ where τ_v is the decay time while moving at velocity v and τ_0 is the decay time while at rest. If it is presumed that decay takes an average number of orbits at rest and that the decay process depends upon the number of orbits (i.e. the number of trials at breaking loose) then it follows that

$$\tau_v/\tau_0 = 1/\sqrt{1-\beta^2}\ . \tag{3.30}$$

This gives the time dilation effect produced by the special theory of relativity but here derived from a classical Newtonian basis.

When any matter particle translates from one place to another along a nominal straight path, it always undulates from one side to the other as it moves. All matter at rest consists of elementary matter particles which consist of mass moving at the speed of light in circular paths. Ordinary matter, such as a bar of steel, is accelerated from rest by photons being transferred, usually from other matter (such as during impact by another bar). In the previous paragraphs we discussed how these photons interact with the flow fields produced by matter to accelerate the matter particles. When a small (low energy-long wavelength) photon interacts with a matter particle the distance from the particle center of mass to the coupling position must be in the order of the wavelength of the photon. Angular momentum considerations require that small impacts be at large distances from the center of the matter particles. Thus, the smaller the interacting photon energy (and the longer the wavelength) and the smaller the resulting matter particle velocity the greater the eccentricity of the coupled mass. Now consider a free wheel in space rotating with a small unbalanced mass placed at a large radius. The axis of the wheel will undulate as it translates to keep the center of mass following an exact straight line. As a result the center of the wheel will take a sinusoidal path to the left and right of the center-of-mass straight path. Let us now calculate the wavelength of the moving matter geometric center path.

Let r_c be the distance from the matter particle's center to the place where the momentum is captured, by both the captured mass and scattered mass. The linear momentum "*P*" times the capture radius is Pr_c - which is assumed to be $\hbar$. For low velocities (i.e. non-relativistic conditions) the momentum *P* is also $M_0\text{v}$, where M_0 is the matter particle rest mass. Now, we can write

$$\hbar = Pr_c = M_o\text{v}r_c, \quad h = 2\pi\hbar = M_o\text{v}2\pi r_c = M_o\text{v}\lambda \qquad (3.31)$$

where λ is the wavelength. Thus

$$\lambda = h/(M_o\text{v}) = h/P \qquad (3.32)$$

This is the relation postulated by deBroglie and is called the "deBroglie wavelength". For high speeds (i.e. for relativistic speeds) consideration must be given to matter particle mass growth, the center of gravity difference, and the matter particle contribution to the angular momentum.

The wavelength λ is measured in meters, the constant h is 6×10^{-34} kilogram-meter ^{2} per second (i.e. Planck's constant), M_0 is the matter particle rest mass in kilograms, and "v" is the particle translational velocity in meters per second. An electron with a mass of $10^{-30}\,kg$ and a velocity $1/3$ the speed of light (i.e., $10^8\,m/s$) has a wavelength of

$$\lambda = \frac{6x10^{-34}}{10^{-30}x10^{8}} = 6x10^{-12} \text{ meters} \tag{3.33}$$

– a very small wavelength. The amplitude of the oscillation is much smaller than the wave length.

The observation of high speed mass growth with velocity is a significant part of high speed (relativity) physics and the observation of matter moving as a wave is a significant part of small item (quantum) physics. Both of these mechanisms come about simply from the interaction of a photon with matter – as shown by the mass growth formula and the wavelength formula just derived.

Throughout the 20th century many authors have stated the impossibility of deriving the special relativity results from classical (Newtonian) theory. We have shown that the three primary relativity observations (mass growth, matter shortening, and time dilation) are derived in a straightforward manner from a classical kinetic particle theory. Further, the mass-energy equivalence $E = M_0 c^2$ is an obvious result of the theory. Finally, we have derived the wave properties of matter rather than postulating them – as done in contemporary physical theory. In summary, these results indicate that the universe is a classical based system.

4. Electrostatics and Magnetism

In this Newtonian universe of hard particles making up an ether, the assemblages of these particles (i.e., neutrinos) orbiting at the speed of light in circular orbits make up all matter at rest. The assemblages making up matter all have angular momentum which is either right-handed or left-handed. The different handedness makes the difference between positive and negative electrostatic charge.

The orbiting assemblage making up a charged particle produces a pulsation in the ether like that of a breathing sphere. Two such assemblages whose centers are at a distance "R" apart can produce an inverse square force of interaction between them. The maximum magnitude of the force produced is given by Bassett[1] [10] as

$$F_e = \rho_0 \frac{8\pi^2 a^2 b^2 \alpha\beta}{T_e^2 R^2} \tag{4.1}$$

This analysis of Bassett is reproduced as Appendix C. In this equation "a" is the nominal radius of one breathing sphere (the proton orbital radius), "b" is the nominal radius of the other, "α "is the half amplitude of oscillation on one sphere which is taken as the proton orbital radius, "β" is the half amplitude of the other sphere, T_e is the period of charge oscillation, ρ_0 is the background mass density, and R is the separation distance. For more background see Whittaker [6]. With electrostatic charges all charges are alike except for the sign which, in this kinetic particle theory here, is controlled by the direction of rotation of the charge-producing assemblage (i.e., the neutrino) about its orbital tangential velocity. Thus, we set $b=a$ and $\beta = \alpha = a$ so that

[1] In Bassett the background density is taken as unity and does not appear in the formula.

$$F_e = \rho_o \frac{8\pi^2 a^6}{T_e^{\,2} R^2} \tag{4.2}$$

The period of oscillation is

$$T_e = 2\pi a / c \tag{4.3}$$

Now

$$F_e = \rho_o \frac{2a^4 c^2}{R^2} \tag{4.4}$$

Since (4.4) gives the electrostatic force, the force also is e^2/R^2 so that

$$e = \sqrt{2\rho_0 a^2 c} \tag{4.5}$$

and using "*a*" as r_p we have (from Appendix A)

$$e = \sqrt{2\rho_0 r_p^{\,2} c} = \sqrt{2\left(1.0495\times10^{19}\right)\left(1.0516\times10^{-16}\right)^2\left(2.9979\times10^{8}\right)} \tag{4.6}$$

$$= 1.5189\times10^{-14}\, kg^{1/2} m^{3/2} s^{-1}$$

Which agrees with the measured value, as it must since the measured value of "*e*" was used to determine the basic constants.

Let us now determine the effect of motion on the electromagnetic force between two charged particles. An electron at rest, in the assumed kinetic particle universe, has an assemblage of kinetic particles orbiting at the speed of light in a circular path. In order to accelerate an electron a photon with angular momentum "$\hbar$" "impacts" the electron electrostatic field. The angular momentum of the combined assemblage (consisting of the electron and the captured portion of the photon) increases by $\hbar$ and the two entities combined translate at velocity "v". The angular momentum then of the combined entities is $\hbar = m r_c \mathrm{v}$, where *m* is the mass (of the two entities) and "r_c" is the half-amplitude of the center of the "charge". The center of mass, of course, continues on a straight path. The undulation of the center of charge is the "electron wave".

Consider now two like electric charges moving at velocity v parallel to each other and with a vector R starting at one charge and ending at the other and which vector is perpendicular to the particle velocities. In a reference frame moving at velocity "v" the two charges are seen to oscillate along the vector v. Assuming phasing is controlled by the twist component of the orbiting assemblage producing the charged particle, the maximum force of interaction between the two particles is given by the same form as the formulas for electrostatic charge, again see [2] and [6]. The difference is that a will be the deBroglie wave amplitude of the charge (which is $\lambda/(2\pi)$), and the period T_m will be λ/v. The force then due to motion will be

$$F_m = \rho_0 \frac{8\pi^2 a^4 \alpha^2}{T_m^{\ 2} R^2} = \rho_0 \frac{8\pi^2 a^4 \left(\lambda/(2\pi)\right)^2}{\left(\lambda/\mathrm{v}\right)^2 R^2} = \rho_0 \frac{2a^4 \mathrm{v}^2}{R^2} \tag{4.7}$$

Dividing the magnetic force by the electrostatic force gives

$$\frac{F_m}{F_e} = \frac{\mathrm{v}^2}{c^2} \tag{4.8}$$

for the special case of equal charges, equal and parallel velocities, and a charge separation vector initiating on one charge and ending on the other where the vector is perpendicular to both velocities. This ratio, of course, is the ratio of the magnetic force to the electrostatic force for this special case. Thus this mechanism models the magnetic force.

If the charges have the same sign then the force is attractive, if opposite, the force is repulsive. By the same mechanism, for which an understanding has not been developed, if the charge velocities are opposite the repulsion/attraction is reversed.

Let us now generalize the special case just developed. Consider a velocity v_1, of charge 1 which produces 100 oscillations in a given period of time. Starting with velocity v_2, of charge 2 with 100 oscillations, if the velocity is reduced say to 1 oscillation in the same time period then the force, clearly, would be reduced to 1/100th of the initial value. Thus, we can generalize the magnetic force equation to

$$F_m = \frac{e^2}{R^2}\left(\frac{v_1}{c}\right)\left(\frac{v_2}{c}\right) \tag{4.9}$$

where v_1 and v_2 can be any value, negative or positive.

The next generalization is for unlike charge magnitudes. We let N_1 be the number of elementary charges at one location and N_2 at the other and take

$$q_1 = N_1 e \text{ and } q_2 = N_2 e \tag{4.10}$$

then

$$F_m = \frac{q_1 q_2}{R^2}\left(\frac{v_1}{c}\right)\left(\frac{v_2}{c}\right) \tag{4.11}$$

If v_1 and v_2 are perpendicular to each other then the force would be zero because of the phasing and should vary sinusoidally from zero when perpendicular to a maximum magnitude when parallel (or anti-parallel).

Finally, if the radius vector *R* for the general case starts at charge 1 and ends at charge 2 (no matter what the relative locations and directions that the charges have) we have the magnetic force given by

$$\underset{\sim}{F}_m = \frac{q_1 q_2}{c^2}\underset{\sim}{v}_2 \times \underset{\sim}{v}_1 \times \frac{\underset{\sim}{i}_R}{R^2} = \frac{q_1 q_2}{c^2 R^2}\underset{\sim}{v}_2 \times \underset{\sim}{v}_1 \times \underset{\sim}{i}_R \tag{4.12}$$

In this expression q_1 and q_2 are the point charges with units of $kg^{1/2}m^{3/2}s^{-1}$, v_1 and v_2 are the charge velocities in *m/s*, $\underset{\sim}{i}_R$ is a unit vector from charge 1 to charge 2, *R* is the magnitude in meters of the vector from charge 1 to charge 2, *c* is the speed of light in *m/s*, and F_m is the magnetic force in newtons, which is attractive in the case where the velocities are parallel and equal and the charges are of like sign.

The above expression can be put in a more familiar form using the concept of the magnetic field. Let the magnetic field generated by charge 1 be

$$\underset{\sim}{B} = \frac{q_1}{c^2 R^2} \underset{\sim}{v}_1 \times \underset{\sim}{i}_R \tag{4.13}$$

Now the magnetic force on charge 2 is

$$\underset{\sim}{F}_m = q_2 \underset{\sim}{v}_2 \times \underset{\sim}{B} \tag{4.14}$$

The electromagnetic units also can be changed to Coulombs and Teslas, if desired.

Let us consider now the effect of relativity. All assemblages making matter when at rest orbit in circular paths. To accelerate a particle from rest, mass is added and the path is changed to a spiral. If the spiral is viewed from a frame moving with the center of mass of the moving particle then the path is elliptic. The time for an orbit is increased as given by the relation

$$\tau_v = \tau_0 / \sqrt{1 - (v/c)^2} \tag{4.15}$$

where τ_v is the period while moving, τ_0 is the period while at rest.

For two charged particles of like charge moving parallel to each other at absolute velocity v and where the vector connecting the two particles is perpendicular to v then the electromagnetic force between them is a force of repulsion given by

$$F_{em} = \frac{q_1 q_2}{R^2} \left[1 - \left(\frac{v}{c} \right)^2 \right] \tag{4.16}$$

If these charges are viewed from a frame moving at the same velocity as the charges then the separating force must be the same as given above. However, when seen in this moving frame the charged particle response would appear to be that due to a force

$$F'_{em} = \frac{q_1 q_2}{R^2} \tag{4.17}$$

since the particle velocities in this frame would be zero.

The particle response is experienced only by the acceleration and in this moving frame it would by $d^2y/d\tau_v{}^2$, if the y-axis is taken to pass through the two particles. The response then as measured by a clock at rest would be

$$\frac{d^2y}{d\tau_v{}^2} = \frac{d^2y}{d\tau_0{}^2}\left(\sqrt{1-(v/c)^2}\right) = \frac{d^2y}{d\tau_0{}^2}\left[1-(v/c)^2\right] \qquad (4.18)$$

Thus, the force would have to be reduced by the factor $\left[1-(v/c)^2\right]$. If the charges are of opposite sign then the electrostatic force is attractive but the magnetic force is repulsive so that the same factor $\left[1-(v/c)^2\right]$, results.

5. Neutrino, Proton, Electron, and Neutron Structures

The basic assemblage of ether particles translates at the velocity $v_r - v_m$, which is the speed of light, has a spin of $1/2$, and has a very small cross section on the order of neutrino cross sections, and thus this assemblage models the neutrinos. The single orbiting neutrino which has an orbital angular momentum of $\hbar/2$ is the proton. The electron is formed when the proton is formed and the electron's formation implies a particular triple-looped structure. Given this structure there is an implied structure for the neutron, which we present. Let us now review the model of the proton which the kinetic particle theory implies. The proton is a neutrino moving at the speed of light in a circular orbit. The proton angular momentum is $\hbar/2 = m_p r_p c$ -- from which the proton path radius r_p is given by

$$r_p = \hbar/(2m_p c) = 1.0545716\times10^{-34}/(2\times1.672621\times10^{-27}\times299792458) \qquad (5.1)$$

$$= 1.051545\times10^{-16}$$

The electrostatic field is produced by the (small) assemblage acting as a ball moving in this circular path – and has been given here previously. The magnetic moment is produced by the circular traveling pulse flow in the first few waves (of the electrostatic field) surrounding the sphere defining the disturbance of the path of radius r_p. To agree with the measured magnetic moment this radius must be $2.79r_p$ -- which seems reasonable but we have not been able to precisely calculate its value. The anti-proton is formed from a translating assemblage with spin opposite that of the one forming the proton.

Let us now discuss the strong nuclear force between two protons (or nucleons). For the time being we will assume that the electrostatic fields of the two protons are non-existent. Of course, we are implying that one of the protons is a neutron, but we will develop the neutron later. We want to discuss the strong nuclear force at this time.

Let us talk about the region between two parallel orbit planes separated by a distance "R" with a proton center at "R/2" from each orbit plane, see Figure 5.1. Let "R" be one-half the mean free path. There will be a general

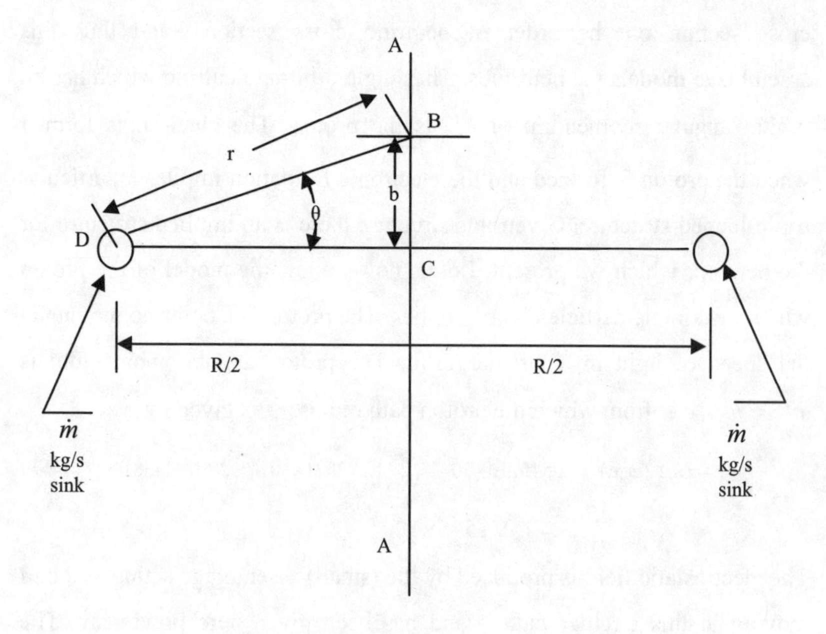

5.1 Hydrodynamic Flow Analysis of Two Sinks

inflow toward the center of the proton at a small area. The basic ether particles, of course, are expelled in a very small area. Now, consider another proton (without an electrostatic field). If it came in the vicinity of this proton, the two would be "pulled" together (actually pushed together by the background pressure being higher than the pressure between the two because of their two "sinks"). They would take parallel orbits with their sinks as close together as possible. The limit on their closeness is established by the density increase as the background particles flow into the sinks. This distance is believed to be in the order of one-third of the mean free path.

Basically for the strong nuclear force we have two flat planes attracting each other and the sinks stay in synchronization as close to each other as possible—which stabilizes the orbits and makes them parallel. The spike's exit flows are like two water hoses spewing out water as they rotate in parallel planes. These "rocket planes" are very thin and parallel, and are many spike cross-section diameters apart—possibly a billion diameters apart.

We can estimate the force between these two such particles using an inviscid fluid. The analysis is further simplified and less accurate, if we ignore density increase—which we will do here. Actually, by considering the density increase, the actual separation distance can be computed and is not significantly different from that resulting from the assumed distance.

We compute the attractive force between two hydrodynamic sinks by determining the flow velocity at the perpendicular plane through the bisection point between the two sinks, i.e., the plane A-A in Figure 5.1. The static pressure on this plane is the ambient pressure less the dynamic pressure, i.e., $p_0 - (1/2)\rho \mathrm{v}^2$. In this expression p_0 is the ambient pressure and v is the flow velocity. The reduction of pressure on this plane, compared with planes far removed from the sink, thus is taken as $(1/2)\rho_0 \mathrm{v}^2$. The attractive force is the integral of this pressure over the plane A-A.

The inflow of fluid is $\dot{m}\, kg/\sec$, (i.e., the sink strength). The mass inflow is $\rho_0 A\mathrm{v}$, where ρ_0 is the density, A is the area, and v is the flow velocity. From mass continuity $\rho_0 A\mathrm{v}$ is constant and, for incompressible flow, Av is constant. Let v_m be the mean background velocity. Let v_s be the value of radius at which the inflow speed is $0.8\mathrm{v}_m$. Now $r_s^2(0.8\mathrm{v}_m) = r^2\mathrm{v}$. Thus $\mathrm{v} = 0.8\mathrm{v}_m r_s^2/r^2$.

The component of this velocity at the plane A-A parallel to the plane is directed from B to C in Figure 5.1, and its magnitude is $\mathrm{v}\sin\theta$.

Due to the right sink, the component also is $v\sin\theta$ so that the total flow is $2v\sin\theta$. There is no flow normal to the plane due to symmetry of the sinks. The pressure reduction now is

$$p = \frac{1}{2}\rho_0 (0.8)^2 \, v_m^{\;2} r_s^{\;4} / r^4 \tag{5.2}$$

and the force is

$$F_n = \int p dA = \frac{1}{2}\rho_0 (0.64) v_m^{\;2} r_s^{\;4} = \int_0^{\pi/2} \sin^2\theta \frac{2\pi b db}{r^4} \tag{5.3}$$

Now, $b = r\sin\theta$, $db = r\cos\theta d\theta + \sin\theta dr$, $r\cos\theta = R/2$, $dr\cos\theta - r\sin\theta d\theta = 0$. Integrating this expression gives

$$F_n = \frac{0.64\pi\rho_0 v_m^{\;2} r_s^{\;4}}{R^2} = \frac{2.01\rho_0 v_m^{\;2} r_s^{\;4}}{R^2} \tag{5.4}$$

Let us compare the equation for the nuclear force with the electrostatic force equation (4.4)

We have

$$F_e = \rho_0 \frac{2a^4 c^2}{R^2}, \qquad F_n = \rho_0 \frac{2.01 r_s^{\;4} v_m^{\;2}}{R^2}, \tag{5.5}$$

There are three significant differences:

1. $c(= v_r - v_m)$ vs. v_m
2. a vs. r_s
3. F_e is due to a sinusoidal flow vs. F_n is due to a steady flow

Everything else being the same the assumed steady flow will produce a greater force than the sinusoidal flow with the maximum amplitude equal to the steady flow velocity. We expect that the flow would be sinusoidal. Further, the values of r_s and "a" are somewhat indefinite. These values could be such that they would compensate for the assumed steady versus sinusoidal flow then F_e to F_n would be in the ratio c^2 to $v_m^{\ 2}$. Experimental results indicate that is what occurs.

Contemporary physics considers the strong nuclear being mediated by resonances with a lifetime of $10^{-23}\,s$ and the proton being a composite of other more fundamental particles. The kinetic particle model of the proton, of course, is just the one orbiting neutrino. This particle will produce organized disturbances in its immediate vicinity. These disturbances will have organized structures and their lifetimes will be in the order of the orbit time. The orbit time "τ" is the proton orbital circumference $\left(2\pi r_p\right)$ divided by the orbit speed "c". Thus

$$\tau = 2\pi r_p/c \doteq 2\pi \times 10^{-16}/\left(3\times 10^{8}\right) \doteq 2\times 10^{-24}\,s \tag{5.6}$$

When the proton is formed it is necessary to form another structure to balance the effect on the background of the proton. The most obvious balancing structure is the one producing the negative electrostatic charge field. The structure also must have angular momentum of $\hbar/2$. Finally, it is necessary that the structure be such that it can have an existence separate from the proton. These properties, of course, are properties of the electron.

Our best model of a structure satisfying the above requirements for the electron is presented now.[1] Figure 5.2 shows the electron structure which is actually just the path taken by extremely small mass (having a cross section in the order of $10^{-46}\,m^2$) [2], making the electron.

---

[1] This model was first published on page 156 of Brown [11] and is shown on the front cover of this reference.

[2] The cross sectional area of the electron (core) is estimated by equating the volume of the electron divided by the volume of the basic particle times the particle mass to the electron mass, which is $\left[\left(4/3\,\pi r_v^3\right)/\left(4/3\,\pi r_b^3\right)\right]m_b = m_e$. The area thus is $A_e = \pi r_v^2\left(m_e/m_b\right)^{2/3} \doteq 10^{-46}\,m^2$. This is in fair agreement with the collision cross sectional area given by (5.15) and the core size given on page 7.

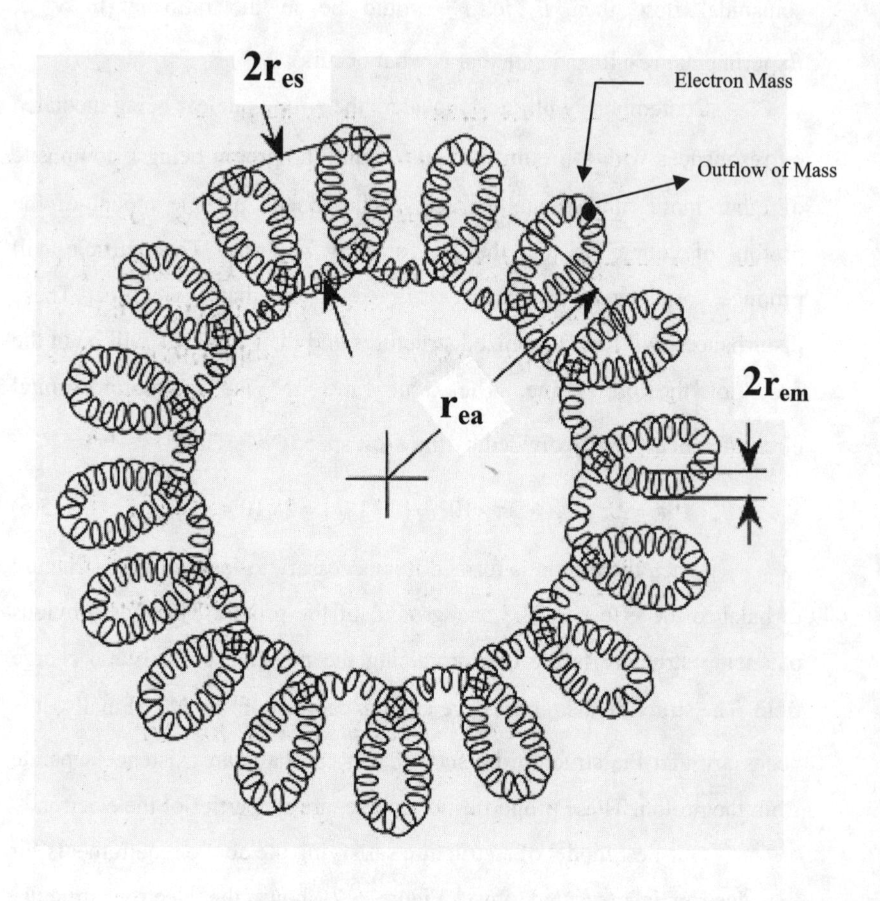

Figure 5.2 Electron Structure

The dimensions on the drawing are r_{em} having a value in the order of $10^{-19} m^2, r_{es}$ in the order of $10^{-16} m$, and r_{ea} in the order of $10^{-13} m$. The smallest loop, with radius r_{em} is the inertial balancing loop where the assemblage large thrust force is balanced by the centrifugal force of the electron mass, the larger loop $(2r_{es})$ is the loop balancing the electrostatic charge of the proton, and the large circular loop is the loop producing the angular momentum of the electron (i.e., the spin of magnitude $\hbar/2$). We present those analyses of the loops which we have been able to develop.

Estimates have been made that the flows involved producing the protons require less mass than the background mass so that upon formation the proton envelope expels a mass equal to the electron mass. This expelled excess mass is formed as one of these organized assemblages and takes a circular path which makes it into matter (just as the translating assemblage above produced a proton). This orbiting assemblage could be thought of as the electron – but there are two other structural components involved to give the electron its properties.

The electron orbital (or "mass") radius r_{em} is controlled by balancing the assemblage thrust (having the same value as the proton assemblage thrust) with the centrifugal force produced by the electron mass m_e. Thus

$$r_{em} = r_p\, m_e / m_p = 1.051545 \times 10^{-16} / 1836.152668 = 5.7268951 \times 10^{-20}\, m \qquad (5.7)$$

Simultaneous with the formation of the small orbital inertial balancing structure a structure must be formed to balance the positive electrostatic field of the proton. The electrostatic field component is produced by a loop of the inertia balancing paths. This field component must travel at the speed of light and this disturbance is produced by the electron making all the inertial loops. The average velocity of the electron as it advances around the electrostatic loop is much less than the speed of light. Thus, r_{es} is considerably less than the proton radius, r_p.

The strength of the field produced by the electron at the radius r_p is the same as that produced by the proton since both fields are caused by an orbiting assemblage whose mass flow rate from input at velocity v_m and flow output rate at velocity v_r are the same for both assemblages.

Since the velocity of propagation of the electrostatic loop around the final circle, the angular momentum circle, is slower than the speed of light. This slower speed would necessitate that the angular momentum radius be greater than the radius required if the electron were moving at the speed of light. This results since the angular momentum is produced by the electron mass traveling in its nominal circular path of radius r_{ea}.

Some additional insight into the electron mechanism can be obtained from the muon. The muon has a mass over 400 times the electron, has the same charge, has angular momentum of $\hbar/2$ (spin 1/2) and has a lifetime of 10^{-6} seconds. Based on the electron model the muon would have 3 structural paths: the inertial loops with radius 400 times the electron inertial radius, the electrostatic loops (same as the electron's), and an angular momentum circle with a radius of $10^{-13}/400 \doteq 10^{-15}\,m.$ The unknowns of the muon, the number of inertial and electrostatic loops and the mass, possibly could be determined from the time-phased action of the muon mass outflow (at the rms velocity) and the inflow (at the mean velocity). The lifetime determination would also have to be determined from these time-phased impulses. In summary, using the basic mechanism of the weak nuclear force, as exemplified by the analysis of the neutron decay, it may be possible to completely determine the masses and structures of the electron and muon.

This model of the electron consisting of a small point mass traversing the inertial loop, these loops traversing the electrostatic loops, and these loops traversing the angular momentum circle has implication for the structure of the neutron.

The neutron, according to the theory here, results when pressure is applied to the hydrogen atom (i.e., photon "mass" is added to accelerate the orbiting electron) so that the angular momentum path is collapsed. The remaining paths then of the electron, the inertial loops and the single electrostatic loop, as an entity would speed up and orbit the proton at close range. The electrostatic force would hold the (modified) electron to the proton. The velocity of the electron is determined by its mass increase (neglecting the smaller mass increase of the proton) which is the neutron mass less the electron and proton masses. The mass increase of the electron is

$$m_n = m_p + \frac{m_e}{\sqrt{1-\beta^2}} \qquad (5.8)$$

Solving for β gives

$$\beta = \sqrt{1-\left[m_e/\left(m_n - m_p\right)\right]^2} = \sqrt{1-\left[9.109\times10^{-31}/\left[\left(1.67493-1.67262\right)\times10^{-27}\right]\right]^2} \qquad (5.9)$$

$$= \sqrt{1-\left(0.394\right)^2} = 0.919$$

The electron mass m_v in the neutron is

$$m_v = \frac{m_e}{\sqrt{1-\beta^2}} = \frac{m_e}{\sqrt{1-\left(0.919\right)^2}} = 3.54 m_e \qquad (5.10)$$

The orbital radius r_n of the reduced electron in the neutron can be determined from

$$F_n = \frac{e^2}{r_n^{\ 2}} = \frac{m_v \mathrm{v}^2}{r_n} \qquad (5.11)$$

or

$$r_n = \frac{e^2}{3.54 m_e \left(\beta c\right)^2} = \frac{\left(1.5\times10^{-14}\right)^2}{3.54\times9.109\times10^{-31}\left(0.919\times3\times10^8\right)^2} \qquad (5.12)$$

$$= 0.92\times10^{-15} m$$

which value is close to the measured size of a neutron. Richtmeyer [12] page 524, gives r_n as $1.4\times10^{-15} m$ as the measured value.

The angular momentum $\hbar/2$ of the neutron is the proton angular momentum combined with the orbital angular momentum of the (partial) electron (which is small) and the spinning angular momentum of the (reduced) electron (which is small).

Neutron decay could result from the random chance that the proton flow output at velocity v_r impacted the electron mass in precisely the way required to push it out to the Bohr radius. The chance of this occurring depends upon the number of tries per second times the half life (in seconds) times the mutual collision area. Let A_c be the mutual collision area and A_b be the Bohr sphere area $\left(=4\pi r_b^{\,2}\right)$. The chance of the required (precise) collision is $A_c/\left(4\pi r_b^{\,2}\right)$. The number of tries produced by the proton in 900 seconds (the neutron lifetime) is

$$900\times c/\left(2\pi r_p\right)=900\times 3\times 10^8/\left(2\pi\times 10^{-16}\right)=4\times 10^{26} \qquad (5.13)$$

From this

$$\frac{4\pi r_b^{\,2}}{A_c}=4\times 10^{26} \qquad (5.14)$$

and

$$A_c=\frac{4\pi\times\left(0.5\times 10^{-10}\right)^2}{4\times 10^{26}}=7.9\times 10^{-47}\,m^2 \qquad (5.15)$$

This is in the order of the cross section area of a neutrino, and agrees fairly well with the electron mass cross section of $10^{-46}\,m^2$, see footnote 2 on page 33 and the footnote on page 7. Thus the above mechanism is concluded to be the cause of the weak nuclear force.

6. Gravitation and the Non-Expanding Universe

The gravitational field is produced by pairs of an electron orbiting a proton in atoms and by pairs of the "reduced" electron orbiting a proton in the neutron. The electrostatic field of a proton begins in the first waves surrounding the proton ("ball") as it orbits at the radius $r_p = 1.052 \times 10^{-16} m$. These closest electrostatic waves are centered on a sphere with a radius of $2r_p$ with its center at the center of the proton. Each wave of the electron electrostatic field (of the orbiting electron in the atom or the reduced electron in the neutron) consists of a flow which is opposite the electrostatic field of the proton except for the "graniness" of the gas producing the field and this "graniness" is controlled by the diameter of the particle making up the gas. This particle causes the two waves (the wave of the proton positive charge and the wave of the electron negative charge) to roll around with respect to each other with a half amplitude equal to the particle radius $r_b \left(\doteq 10^{-35} m \right)$. In effect this acts as an enveloping sphere of radius $3r_p$ breathing with a half amplitude of the basic particle radius r_b. With these changes the field is like the electrostatic field and the force of interaction between one (+ and -) charge pair and another (+ and -) charge pair separated by a distance of R can be written. With the above modifications to Eq. (4.1) the gravitational force is given as

$$F_g = \rho_o \frac{8\pi^2 \left(3r_p\right)^4 r_b^2}{\left(2\pi r_p / c\right)^2 R^2} = \frac{162 \rho_0 r_p^{\,2} r_b^{\,2} c^2}{R^2} = \mathrm{G} \frac{m_g^{\,2}}{R^2} \tag{6.1}$$

where m_g is the average mass of neutrons and the proton-electron pairs in atoms and G is the universal gravitational constant. Solving for G and substituting the values for the parameters and m_p for m_g gives

$$G = 162\rho_0 r_p^{\ 2} r_b^{\ 2} c^2 / m_g^{\ 2} \tag{6.2}$$

$$= 162\left(1.0495\times10^{19}\right)\left(1.0516\times10^{-16}\right)^2\left(1.0516\times10^{-35}\right)^2\left(2.9979\times10^{8}\right)^2/\left(1.6726\times10^{-27}\right)^2$$

$$= 6.672\times10^{-11}\, m^3 / kg - s^2$$

Which agrees closely with the measured value

$$G_{meas} = 6.673\,(10)\times10^{-11}\, m^3 / kg - s^2 \tag{6.3}$$

as expected since this value of G was used to determine the basic constants.

The "graniness" of the electrostatic field is also believed to be the cause of photon decay.[1] As the photon goes through each cycle of its translational path it apparently loses exactly one basic ether particle by a mechanism which has not yet been identified. Assuming that this is the case then photon dissipation would produce the same observable as an expanding universe. The following analysis shows how this occurs.

According to the expanding universe theory the velocity "v" of each distant star measured relative to the earth is directed from the earth to the star and its magnitude is

$$\mathrm{v} = H\, r \tag{6.4}$$

where H is the Hubble constant $\left(=1.80\times10^{-18}/s\right)$, and "r" is the distance form the earth to the star. The velocity is related to (the observed) wave length λ_{obs} by

$$\mathrm{v} = Hr = c\frac{\lambda_{obs} - \lambda_o}{\lambda_o} \tag{6.5}$$

where λ_o is the emitted wave length. The analysis giving this equation is limited to λ_{obs} being less than $2\lambda_o$, since a value greater than that would

[1] This analysis grew out of a dimensional analysis interrelating microscopic, macroscopic, and cosmological parameters initiated by Dr. Robert M. Wood.

correspond to a star receding at a velocity greater than the speed of light. In that case no waves would reach the earth, according to the Hubble theory.

Let us now examine the consequences resulting from the indicated loss of one ether particle of mass m_b from a photon for each wave length of its travel. The amount of "mass" loss per meter of the travel is

$$\frac{\Delta m}{\Delta r} = -\frac{m_b}{\lambda} \tag{6.6}$$

where the negative sign indicates the reduction in mass and λ is the photon wavelength at the time the particle is removed. Initially m_b is very small (30 orders of magnitude less than m) so that we can set $\Delta m / \Delta r$ equal to dm / dr. Now since $E = hc / \lambda = mc^2$ we have

$$\frac{dm}{dr} = -\frac{m_b}{\lambda} = -\frac{m_b mc}{h} \tag{6.7}$$

Integrating gives

$$\int_{m_0}^{m} \frac{dm}{m} = -\frac{m_b c}{h} \int_0^r dr = \ln \frac{m}{m_o} = -\frac{m_b c}{h} r \tag{6.8}$$

Solving for r gives

$$r = -\frac{h}{m_b c} \ln \frac{m}{m_o} \tag{6.9}$$

Letting m_o be the "mass" of the highest energy photon emitted from a hydrogen atom we have

$$m_o c^2 = E_0 = E_2 - E_1 = 1/2\, \alpha c^2 m_e \left(1/4 - 1\right) = -2.243 \times 10^{-16} J \tag{6.10}$$

Thus

$$m_o = 2.243 \times 10^{-16} \big/ \left(3 \times 10^8\right)^2 = 2.492 \times 10^{-33} kg \tag{6.11}$$

Using $m_b = 1.691 \times 10^{-66} kg$ gives

$$r = -\frac{1.054\times10^{-34}}{1.691\times10^{-66}\times3\times10^{8}}\ln\frac{m}{2.492\times10^{-33}}$$

$$= -2.08\times10^{23}\ln\frac{m}{2.492\times10^{-33}} \tag{6.12}$$

The following table gives values of r and wave length for several values of m/m_o.

$$\left(\lambda = (h/m_o)(m_o/m) = \left(1.054\times10^{-34}\right)/\left(2.492\times10^{-33}\right)(m_o/m) = 0.0422\times(m_o/m)\right) \tag{6.13}$$

m/m_0	0.9	0.5	0.1	0.0001	$m_b/m_0 = 6.79\times10^{-34}$
$r-m$	2.19×10^{22}	1.44×10^{23}	4.78×10^{23}	1.92×10^{24}	1.59×10^{25}
$\lambda-m$	0.047	0.0844	4.22	422	6.22×10^{31}

The value of $r = 1.59\times10^{25}$ is when the photon is at its minimum mass and thus is the maximum range for the hydrogen emitted photon. However, near the end of the range the analysis is not accurate due to the large percentage change in photon mass – as exemplified by the wave length of 10^{32} meters for the last photon wave. The value of the range computed $\left(i.e.\ 1.59\times10^{25}\,m\right)$ compares favorably with the radius of the observable universe of $10^{10}\ ly$ since the radius in meters is

$$r_u = 10^{10}\,ly\times365.25\,d/y\times24\,h/d\times3600s/h\times3\times10^{8}\,m/s \tag{6.14}$$

$$= 9.47\times10^{25}\,m$$

Let us determine the values of range and "velocity" for small values of range, and then determine the predicted value of the Hubble constant, based on this theory. Let us use $m/m_0 = 0.9$. For this value the range is $2.19\times10^{22}\,m$ (see the forgoing table) and the "velocity" based on assuming the wave shift is due to a receding source (i.e. a Doppler shift) rather than the mass loss is

$$\mathrm{v} = c\frac{\lambda_{obs} - \lambda_0}{\lambda_0} = c\frac{\dfrac{h}{m_{obs}c} - \dfrac{h}{m_0 c}}{\dfrac{h}{m_0 c}} = c\left(\frac{1}{0.9} - 1\right) = 0.111c \tag{6.15}$$

Thus

$$H = \frac{\mathrm{v}}{r} = \frac{0.111 \times 3 \times 10^8}{2.19 \times 10^{22}} = 1.52 \times 10^{-15} s^{-1} \tag{6.16}$$

This is a thousand times larger than the measured value based on the Doppler shift theory, $\left(H_{Doppler} = 1.80 \times 10^{-18} s^{-1}\right)$.

In this kinetic particle theory of the photon the wave length shift can be tens of orders of magnitude larger than the emitted wave length and the wave length shift equation for the expanding universe assumption is valid only for a shift of twice the emitted wave length (i.e., for a star at a distance having a velocity of the speed of light). Nonetheless the predictions of the radius of the observable universe for both theories are in fair agreement.

7. Quantum Mechanics

The fundamental nature of this kinetic particle theory of physics clearly provides the foundation of quantum mechanics. The mechanism of the neutrino produces a single value of angular momentum for the neutrino and provides the mechanism for making all fundamental processes move at the speed of light. This latter mechanism, of course, is the fundamental factor in relativity observables. The former mechanism, quantization of angular momentum, is fundamental to all quantum effects.

The neutrino angular momentum is produced by mass rotating about an axis. However, it is not the same as matter rotating about an axis since it is, in effect, a "gaseous" rotating phenomenon. Also, independent of how much mass is in the neutrino core, all neutrinos have the same angular momentum (and particle in-flow rate).

As the neutrino whose mass is that value of mass in a proton gets into circular orbit, it clearly will "stir" the background to produce waves. The primary component of the wave pattern produced is the electrostatic field. This is a field that is almost spherically symmetric. It begins approximately at the proton radius $\left(\approx 10^{-16} m\right)$ and extends until it fades into the background ether.

Basically, the proton field is a spherical wave in a gaseous ether which has an electrostatic potential. The mass of the proton is all concentrated in a very small region of the proton. It is not possible to measure (or determine) its position within the accuracy of the diameter of the basic ether particles—at least with the techniques available from the mainstream theory of physical science.

What we would like to do here is to show the basis for the "wave equation" representing the hydrogen atom. We will concentrate on understanding the Schroedinger equation for the hydrogen atom without considering the effect of electron spin or relativistic effects.

First, we derive the deBroglie wave equation for a moving electron. We then derive the Schroedinger equation for a hydrogen atom.

Louis deBroglie apparently reasoned that since Albert Einstein and Max Planck had shown that light with its wave behavior had particle behavior, matter, likewise, with its particle properties should have wave properties. Basically he postulated that matter should translate in an undulatory manner with a wave length given by

$$\lambda = \frac{h}{m\mathrm{v}} \tag{7.1}$$

where "m" is the mass of the particle, "v" is its velocity, and h is Planck's constant $(h = 2\pi\hbar)$.[1] In this present theory we derive, instead of postulate Eq.(7.1).

Recall that the structure of an electron consists of a very small neutrino orbiting in a very small path (10^{-20} m in diameter). This path is a perturbation to the larger path (10^3 times the diameter of the neutrino orbital path) which produces the electrostatic field. This larger path is a perturbation to the even larger path (10^3 times the diameter of the electrostatic field path) which is the path producing the electron angular momentum. It turns out that the wave property of an electron, as represented by the deBroglie relation and as used in the Schroedinger equation, is a very simple part of the complex structure of the electron field.

In order to understand the origin of the deBroglie relation, we need to recall how an electron has its velocity changed. Imagine an electron at rest (knowing full well that an electron can never be at rest), which means that the center of the angular momentum field is at rest. Now, impact the electron by a low energy photon. The electron captures the photon and the

[1]Apparently, deBroglie thought of the electron being embedded in a "pilot" wave, which is similar to the theory presented here.

two entities become one entity which is translating at velocity v.[1] The angular momentum of the photon before capture was $\hbar$. The angular momentum increase resulting to the electron, i.e., for the two combined entities is $\hbar = mr\mathrm{v}$, where r is the half-amplitude of the electron wave. Since the photon always moves at velocity *c*, it must be captured at a great distance from the angular momentum center. Linear momentum conservation also shows that the radius must be very large since the captured photon still continues moving at velocity c and the electron photon assembly moves only at velocity v, where v << *c*. The frequency of the wave, then, is $\mathrm{v} = c/2\pi r = c/\lambda$. Using $r = \lambda/2\pi$ gives

$$h = m(\lambda/2\pi)\mathrm{v} \tag{7.2}$$

or

$$h = 2\pi\hbar = m\mathrm{v}\lambda \tag{7.3}$$

and

$$\lambda = h/(m\mathrm{v}) = h/p \tag{7.4}$$

where p is the matter particle momentum. This, of course, is the deBroglie relation.

The extension of this equation to the relativistic case, of course, requires accounting for the mass actually added to the electron.

Before going on to the Schroedinger equation, we should mention the so-called double slit experiment. When electrons pass through a screen with a single slit, they produce a Gaussian type dispersal pattern on a "screen." This pattern results since the relative position of the wave phase and the slit vary from one electron to the next. When the electron goes through the slit, all of its field is squeezed down and goes through that single hole. When a second slit is opened and if the electrons still pass through the initial slit, there is a diffraction pattern on the screen. The

[1] The portion of the electron field which captures the photon is the electrostatic field component, and not the larger radius angular momentum field nor the smaller force balance field.

diffraction pattern results because of the wave phase and slit position varying and because of part of the electron field passing through the second slit.

The key to this phenomenon is that the electron itself is a very small item, and the field is a very extensive item. It is anticipated that the pattern which results on the screen for electrons actually passing through one slit would differ from, but be symmetric with the pattern for electrons passing through the other slit. Incidentally, photons from lasers, for example, should be directable to pass through a pre-selected slit and thus show the different, but symmetric, patterns for the two slits. The electron is a very small particle, but it has a large energy since its energy is

$$m_e c^2 = 10^{-30}\left(3\times10^{8}\right)^2 = 10^{-13} \text{ joules.} \tag{7.5}$$

This energy of the moving mass also is the same as the energy of motion of its field, i.e., the energy stirred up in the background. These dynamic effects produced by the motion in the background are what we observe. When we observe its effects, we say we observe the electron. In this sense we can talk about the electron at a given time being distributed over a region of space. The energy per unit volume varies with time and location since the energy is a wave. This wave has the general equation

$$\psi(x,t) = \psi_o e^{i(kx-\omega t)} \tag{7.6}$$

which is the wave generated by an electron translating along an advancing, but sinusoidal, path. Even though we have used the familiar wave function symbol ψ, do not assume it is necessarily the function ψ used by quantum electrodynamicists. We hope here to define this ψ strictly from a classical Newtonian basis. We will return to this shortly.

The conservation of energy for the electron is

$$p^2/(2m) + P = E \tag{7.7}$$

where E is the electron's total energy, "P" is the electron's potential energy, and "p" is its momentum (p = mv). Thus, $p^2/(2m) = mv^2/2$ is the electron's kinetic energy.

If we want to predict anything about the detailed structure of the electron, i.e., of the electron field produced by the elusive neutrinos, then we can multiply each of the above three terms by a distribution function of space and time which, when evaluated, will give the characteristics of this field for any location and time. We thus have

$$\left(p^2/2m\right)\psi + P\psi = E\psi \tag{7.8}$$

Using the deBroglie relation for all matter particles, we have

$$\lambda = h/p \tag{7.9}$$

or

$$p = h/\lambda = (h/2\pi)(2\pi/x) = \hbar k \tag{7.10}$$

where k is the space variation frequency in the wave equation. Using the Einstein relation (which we have derived classically from this theory), we have

$$E = h/\lambda = (h/2\pi)(2\pi\lambda) = \hbar\omega \tag{7.11}$$

The energy equation can be written as

$$\left[\hbar^2 k^2/(2m)\right]\psi + P\psi = \hbar\omega\psi \tag{7.12}$$

Noting that

$$\partial\psi/\partial x = ik\psi\,\partial\psi/\partial t = i\omega\psi \tag{7.13}$$

we can write this energy equation as

$$-\left(h^2/2m\right)\left(\partial^2\psi/\partial x^2\right) = P\psi = i\hbar\partial\psi/\partial t \tag{7.14}$$

This, of course, is the Schroedinger time dependent wave equation for one dimension.

The function ψ can be a velocity (as well as a square root of energy) function. The function ψ can be defined in such a way that $|\psi|^2$ gives the energy density function. The function ψ is called the probability amplitude. The energy E contained in a volume Δ V at a given point is $E\psi\ \psi * \Delta$ V. The electron energy is

$$\int (E/V)\psi\psi dV = (E/V)\int \psi\psi * dV = mc^2 . \tag{7.15}$$

Further, note that

$$(1/V)\int \psi\psi * dV = 1.0 , \tag{7.16}$$

which is the usual definition of a normalized distribution function.

We note here that ψ itself is similar to a planar velocity since $\psi\ \psi$ * is like the square of two velocity components[1] so that $\psi\ \psi$ * is similar to an energy. Thus ψ is a space and time distribution function of the electron field energy. The occurrence of an "i" on the right side of the Schroedinger equation indicates a 90-degree shift of phase for the velocity. The occurrence of the second partial of ψ with respect to distance and the first partial derivative with respect to time, as well as the phase shift, are the direct result of a classical wave function satisfying the energy equation.

In the early years of development of quantum mechanics, there were two interpretations of the wave function ψ . One was that the electron somehow was spread out over space and ψ was its mass distribution function which, of course, is the interpretation here. It is emphasized that this is a classical Newtonian interpretation when the electron field is being considered instead of the electron itself. However, physicists in the 1920s were not considering the electron as being a flow of classic kinetic particles.

---

[1] Recall that for two complex number A and B, where A = r + is and B = r – is, that AB* is $r^2 + s^2$

The second interpretation of $\psi\ \psi$ *, due to Max Born in the 1920s, is that it is the probability density function such that $\psi\ \psi^* \Delta V/V$ gives the probability of finding the electron in the particular differential volume ΔV_1 (more precisely, in the volume x_1 to $x_1+\Delta x$, y_1 to $y_1+\Delta y$, z_1 to $z_1+\Delta z$ where $\psi\ \psi$ * is evaluated at $x_1\ y_1\ z_1$). It appears that quantum physicists took a wrong turn there by accepting the Born interpretation. However, without the field description developed in this present theory, the first interpretation seemed very bizarre.

Let us continue with our interpretation of ψ. The space and time variables are easy to separate. We write

$$\psi(x,t) = Ae^{i(kx-\omega t)} = Ae^{iky} \cdot e^{-i\omega t} = \varphi(x)\cdot f(t) \tag{7.17}$$

Substituting into the one dimensional wave equation gives

$$-\left(h^2/2m\right)\partial^2\psi/\partial x^2 \cdot f(t) + P(x)\psi(x)\cdot f(t) \tag{7.18}$$
$$= i\hbar\psi(x)\cdot\partial f/\partial t$$

Dividing each term by $\psi(x)\cdot f(t)$

gives

$$-\left(h^2/2m\right)\left(\partial^2\psi/\partial x^2\right)\left(1/\psi(x)\right) + P(x) \tag{7.19}$$
$$= i\hbar\left(\partial f/\partial t\right)/f(t)$$

The left side is a function of x only, and thus equals a constant, or zero. The right side is equal to the negative of the constant. We let the constant be unity for the previous interpretation of ψ. However, without loss of generality and to be in line with current practice, we redefine ψ so that $\int \psi\psi^* dV$ is equal to the total energy of the particle E. Thus we will take the separation constant to be E. Now, the two equations are total differential equations, and are

$$df/dt = -(iE/\hbar)f(t) \tag{7.20}$$

for the right side and for the left side

$$-(\hbar^2/2m)d^2\psi/dx^2 + P(x)\bullet\psi(x) = E\psi(x) \tag{7.21}$$

The solution for the first equation is

$$f(t) = Ce^{-i\omega t} \tag{7.22}$$

For solving the second equation, we write it as

$$\left(-(\hbar^2/2m)d^2/dx^2 + P\right)\psi = E\psi \tag{7.23}$$

or

$$Q\psi = E\psi \tag{7.24}$$

where we define the operator

$$Q = -(\hbar^2/2m)(d^2/dx^2) + P \tag{7.25}$$

We define ψ then over the complete region of the electron and then $E\psi\psi*/\mathrm{V}$ gives the distribution function of the electron field energy as a function of the space coordinates (x, y, z) and time. The integral over the volume is the total energy E. This energy divided by c^2 then is the electron mass.

The undulatory path of the electron gives the quantitization of orbits for the atoms as postulated by Bohr and as shown by deBroglie with his postulated waves.

Here we have shown the classical basis of the electron wave. Knowing that, we have a wave distribution of the electron and we have used a wave energy distribution function with the electron conservation of energy equation and have shown that gives the Schroedinger equation. We are now in a position, based strictly on a classical theory, to write the Schroedinger equation for the hydrogen atom.

The three-dimensional time-dependent Schroedinger equation is

$$-\frac{\hbar}{2m}\nabla^2\psi(x,y,z,t) + P(x,y,z)\psi(x,y,x,t) = i\hbar\partial\psi(x,y,z,t)/\partial t \tag{7.26}$$

where

$$\nabla^2 = \frac{\partial^2}{\partial x^2} + \frac{\partial^2}{\partial y^2} + \frac{\partial^2}{\partial z^2} \tag{7.27}$$

We write

$$\psi(x, y, z, t) = \varphi(x, y, z) e^{i\omega t} \tag{7.28}$$

and substituting into the right side of the general Schroedinger equation gives

$$-\frac{\hbar}{2m}\nabla^2\psi(x,y,z,t) + \rho(x,y,z)\psi(x,y,z,t) = i\hbar(-i\omega)\varphi(x,y,z)e^{-i\omega t} \tag{7.29}$$

$$= \hbar\omega\varphi(x,y,z)e^{-i\omega t} = E\varphi(x,y,z)e^{-i\omega t}$$

where $E = \hbar\omega$. Thus we have

$$-\frac{\hbar}{2m}\nabla^2\varphi(x,y,z)e^{-i\omega t} + P(x,y,z)\varphi(x,y,z)e^{-i\omega t} = E\varphi(x,y,z)e^{-i\omega t} \tag{7.30}$$

Canceling out the $e^{-i\omega t}$ gives the time-independent Schroedinger equation

$$-\frac{\hbar}{2m}\nabla^2\varphi(x,y,z) + P(x,y,z)\varphi(x,y,z) = E\varphi(x,y,z) \tag{7.31}$$

8. Quantum Electrodynamics

a. The Electron Two Slit Experiment

The kinetic particle theory of physics clearly provides the structure underlying all of quantum theory. Thus, any observation explained by quantum electrodynamics must be explained by the kinetic particle theory of physics. One of the most interesting (and, mysterious) phenomenon is the "double-slit" experiment. We show here how the kinetic particle theory explains the "double-slit" experiment for electrons.

Figure 8.1 shows an electron emitter, a "target" with two slits, and a detector screen. Figure 8.2 shoes the two slits and electron hit densities for the three cases of right slit open, left slit open, and both slits open.

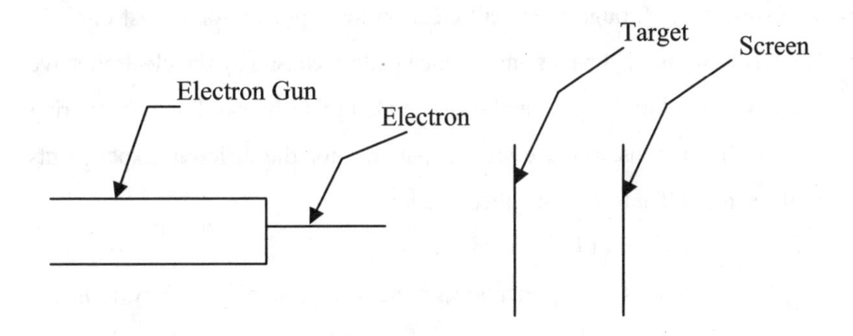

Figure 8.1. Double-Slit Experiment

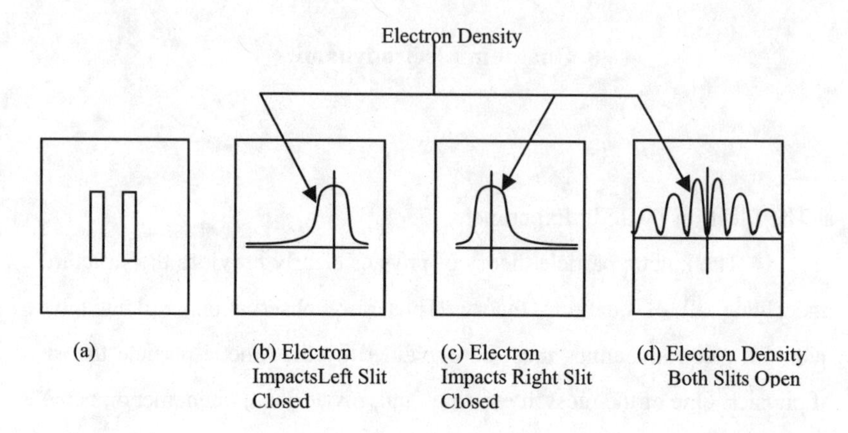

Figure 8.2. Target and Electron Densities for the Three Experiments

The patterns in (b) and (c) are clearly due to the electron wave front hitting the slit at different distances along the wave. The field structure of the electron extends much beyond both slits. Thus any electron impacting the target which passes through the target will have its field pass through both slits, even through the small electron mass goes through just one slit. The result of this will be an interference pattern caused by the electron wave hitting at a random longitudinal position along its wave and then interfering with itself. Photons produce similar patterns for the different experiments by the same self-interference phenomenon.

b. Partial Reflection of Photons by Glass

An interesting application of the kinetic particle theory is in the analysis of partial reflection of light by glass. In the analysis here we consider the same problem analyzed by Feynman [7] in Chapter 1 and 2. If the glass is very thick and uneven on the back surface, such as a lake of water, then 4% of the photons are reflected, even when one photon at a time impinges on the glass. It is believed that this is the light reflecting from the front surfaces of the glass. For glass with plane front and back surfaces the percent of reflection (at least by many photons) varies with thickness as shown by Figure 8.3.

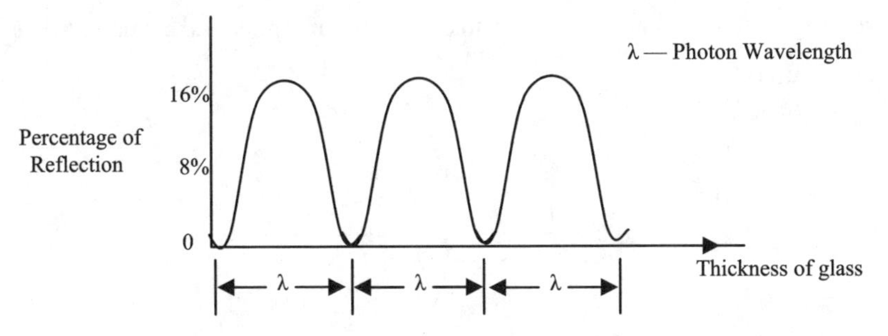

Figure 8.3. Variation of Percent Reflection by Glass as a Function of Thickness

Our analysis of this phenomenon follows.

The photon, in the kinetic particle theory, is a narrow "string" of some 10^{20} to 10^{30} brutinos which extends in a sinusoidal path one wave length. The photon in this analysis is a photon of red light which is 6.5×10^{-7} meters long. The glass is made up of atoms with diameters which are approximately $1/1000\,\text{th}$ as large as the photon. Imagine a plane of glass with the thickness of just one atom. Most of the photons impinging on the glass will pass through. However, 8% will interact with the glass. There are two paths a photon can take. It can couple with the electrostatic fields of the atoms, which are rotating at the same speed as the photon (i.e., at the speed of light) and be deflected (reflected from the front surface) or wrap around the electrostatic field and pass back toward the source. What happens is that the photon takes both paths, part is reflected from the front surface and part is reflected from the back surface.

The photon velocity of the front path can be represented by ψ_1 and of the back path by ψ_2 where ψ_1 and ψ_2 are complex numbers (and thus represent vectors). These functions are called probability amplitudes. The probability of each path is the same but the wave for path 2 lags that of path

1 by the slight time required to go around the atom. Thus, the direction of the two vectors are slightly different and the "senses" are almost opposite, see Figure 8.4. What happens is that the photon splits and actually takes both paths.

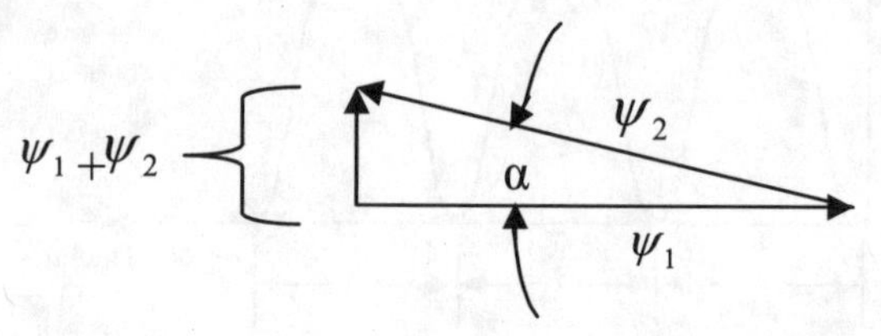

Figure 8.4 Velocities of Reflecting Photon For the Two Paths

For path 2 after the photon portion turns 180º (which the photon travels the distance $(2\pi/1000)\lambda$ inside the glass) and its flow velocity is opposite that of the incoming part. "Self Interference" thus cancels the photon, most of the time. The photon energy is proportional to the square of the velocity. Thus $|\psi_1 + \psi_2|^2$ is the probability of the photon being reflected by the front and back surface. Clearly $|\psi_1 + \psi_2|^2$ is very small compared to $|\psi_1|^2$ and $|\psi_2|^2$ since the glass thickness is so small compared to the photon "length".

As the glass becomes thicker the angle α , in Figure 8.4, becomes larger and larger reaching π radians when the glass thickness is $\lambda/2$. At that thickness the percentage of reflection is 16%, 12% from the back surface and 4% from the front surface. At one atom thickness we say the front surface reflects 4% and the back surface reflects 4% of the photons, but the two are almost exactly out of phase so that practically no photons are reflected.

Figure 8.4 shows the physical significance of the arrows used in quantum electrodynamics. The arrows are used to determine the phasings of the portions of a photon.

c. Diffraction of Light

Consider a light source "s" a mirror, M; and a detector "D" as shown in Figure 8.5. A photon emitted from the source can be reflected in any direction over the 2π sterradians of angle covering the hemisphere about the mirror. The probability of the reflected photon intercepting the detector is the solid angle β divided by 2π.

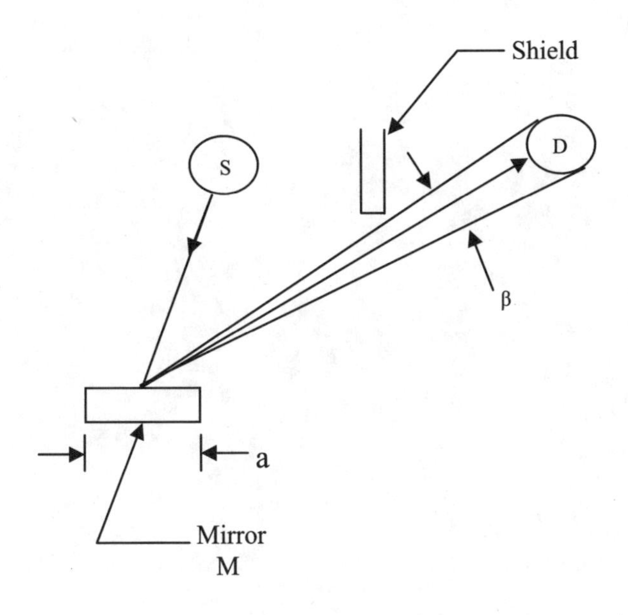

Figure 8.5 Diffraction Experiment

If we add another section of mirror of width "a" adjacent to the right side of the initial mirror then, for a particular value of "a", the probability of a photon reaching the detector decreases to zero. The reason for this is the photon heading for the detector splits and takes two paths inside the mirror, one path of which is close to a half wave length longer than the other path. The photon then "interferes" with itself and cancels itself. A photon impacting mirror sections close to the plane bisecting the source and detector line can have multiple paths but the path lengths are approximately equal. Thus, the photon does not interfere with itself. By proper placing of mirror sections only at the extreme left in Figure 8.4 the detector can "see" the source. Such placement of mirror sections produces a diffraction grating.

9. Closure

The kinetic particle theory of physics provides the following:

1. The fine structure constant is the ratio of force due to the root mean square velocity less the mean velocity, the difference divided by the force due to the mean velocity of the background ether.
2. All matter at rest is made up of kinetic particles moving in a circular path at the speed of light so that the energy of all matter is its mass times the square of the speed of light.
3. A proton is made of a single small orbiting assemblage of background particles.
4. Mass is added to matter to accelerate it so that the mass at velocity is the rest mass divided by $\sqrt{1-\beta^2}$, where β is the matter velocity in speed of light units.
5. When mass is added to accelerate matter the mass is added eccentrically so that the mass is still moving at the speed of light but taking a spiral path. The spiral path takes longer to cycle than when at rest (giving time dilation) and shortens the matter when viewed from a frame translating with the matter.
6. The eccentric mass causes mass to undulate as it moves then this gives matter its wave property.
7. The mass making up a proton or an electron moving in a circular path sets up the electromagnetic field which produces electrostatic forces.
8. The translating wave produces magnetism by the same mechanism that the rotating mass causes electrostatic forces.
9. The electron structure consists of a single concentrated mass taking a path in a loop which balances the inertial forces, a larger loop which produces the electrostatic field, and a third even larger complete circle loop that produces the electron angular momentum.

10. The neutron results when a sufficiently large pressure is applied to a hydrogen atom so that the angular momentum loop of the electron is collapsed and centrifugal force is balanced by the strong nuclear force when the collapsed electron orbits the proton.
11. The weak nuclear force is due to the strong nuclear force less the result of the chance encounter with the output from a basic matter particle assembly where the output mass is moving at the rms velocity of the background ether.
12. Gravitation is caused by two orbiting balanced opposite charges except for that prohibited by the actual basic ether particle size making up all the contents of the universe.
13. The Hubble red shift of star light is caused by photons losing one basic ether particle for each wave length of travel, and giving the illusion of an expanding universe.
14. The kinetic particle flows producing the fields of particles are modeled using the equations of quantum mechanics.
15. Kinetic particle theory explains the double-slit experiments for electrons and photons as well as partial reflection and diffraction grating experiment.

Appendix A. Determination of the Basic Constants of Physics

A universe containing only a gas of inert hard elastic spherical particles of one size requires four, and only four, parameters to completely define it. The four we have selected here are the mean speed of the particle (v_m) (measured relative to a frame for which there is no net flow of gas, i.e., a rest frame), the mass of the individual particle (m_b), the radius of the particle (r_b), and the mean free path for the gas (l). It is obvious that the first three parameters are required, they give the properties of mass, length, and time to the theory. However, if two universes were imagined with the same particle mass, diameter, and speed but a different mean free path then the two universes would be different. For example if the first universe had a mean free path of 4 particles radii and the other a mean free path of 4,000,000 radii, they certainly would be different universes.

If the universe actually were a kinetic particle universe and if we understood the mechanisms by which everything functioned then we should be able to find four independent measured constants for which the four basic constants of this theory could be determined. As it so happens we have an approximate understanding of four different (independent) mechanisms which provide an exact value (within measurement error) for one constant and approximate values for the other three.

The measured constants we use are the speed of light, the value of the basic electron charge, the mass of the proton, and the universal gravitational constant. The mechanisms we use are the one controlling the speed of light for the neutrino, the mechanism producing the electrostatic charge, the mechanism producing gravity, and the mechanism of the neutrino. The measured constants we use are c (the speed of light), e (the charge of the electron), m_p (the mass of the proton), and G (the universal gravitational constant). The mechanisms are all explained in the text and here we will just present the equations. We will generally calculate to 5 significant figures since this is greater than the accuracy of all the mechanisms except the speed of light which gives "v_m" as accurate as "c".

The speed of light computation.

$$c = v_r - v_m,\quad v_r / v_m = \sqrt{3\pi/8},\quad c = \left(\sqrt{3\pi/8} - 1\right) v_m \tag{A1}$$

$$v_m = c / \left(\sqrt{3\pi/8} - 1\right) = 2.9979\times10^8 / 0.085402 = 3.5103\times10^9\ m/s$$

The electrostatic charge computation.

$$F_e = \frac{e^2}{R^2} = \rho_0 \frac{2r_p{}^4 c^2}{R^2},\quad \rho_0 = \frac{e^2}{2r_p{}^4 c^2},\quad \frac{\hbar}{2} = m_p c r_p$$

$$\alpha = \frac{e^2}{\hbar c},\quad r_p = \frac{\hbar}{2m_p c} = \frac{e^2}{2\alpha c^2 m_p}, \tag{A2}$$

$$\rho_o = \frac{e^2 16\alpha^4 c^8 m_p{}^4}{2c^2 e^8} = 8\left(\frac{\alpha^2 c^3 m_p{}^2}{e^3}\right)^2$$

The gravitational force computation.[1]

$$F_g = \frac{162\rho_0 r_p{}^2 r_b{}^2 c^2}{R^2} = G\frac{m_g{}^2}{R^2} \tag{A3}$$

$$r_b = \sqrt{\frac{G}{162\rho_0}}\frac{m_g}{r_p c} = \sqrt{\frac{G}{162(8)}}\frac{e^3 m_g 2\alpha c^2 m_p}{\alpha^2 c^3 m_p{}^2 c e^2} = \sqrt{\frac{G}{324}}\frac{e}{\alpha c^2}\frac{m_g}{m_p}$$

The angular momentum computation.

---

[1] The equation for r_b can be written in terms of the Planck length.

$$\frac{\hbar}{2} = \rho_0 4\pi l^2 \mathrm{v}_m \frac{\pi l}{\mathrm{v}_m}(0.8l)\,\mathrm{v}_m$$

$$l^4 = \frac{\hbar}{63.165\mathrm{v}_m\rho_0} = \frac{1}{63.165}\left(\frac{e^2}{\alpha c}\right)\frac{\left(\sqrt{3\pi/8}-1\right)}{c}\frac{e^6}{8\left(\alpha^2c^3m_p{}^2\right)^2} \tag{A4}$$

$$= 0.00016901\frac{e^8}{\alpha^5c^8m_p{}^4}$$

$$l = 0.11402\frac{e^2}{\alpha^{5/4}c^2m_p}$$

Obtaining m_b from ρ_0 and l.

$$\eta = \frac{1}{\sqrt{2}\pi 4r_b{}^2l}, \quad \rho_0 = m_b\eta = \frac{m_b}{\sqrt{2}\pi 4r_b{}^2l}$$

$$m_b = \sqrt{2}\pi 4r_b^2 l\rho_0 = \sqrt{2}\pi 4\frac{G}{324}\frac{e^2}{\alpha^2c^4}\left(\frac{m_g}{m_p}\right)^2 \times 0.11402\frac{e^2}{\alpha^{5/4}c^2m_p}8\left(\frac{\alpha^2c^3m_p^2}{e^3}\right)^2 \tag{A5}$$

$$= 0.05003\frac{G}{e^2}\alpha^{3/4}\left(\frac{m_g}{m_p}\right)^2 m_p{}^3$$

The above calculations give the four kinetic particle constants in terms of the four measured constants as:

$$\mathrm{v}_m = c/\left(\sqrt{3\pi/8}\right)-1 = 3.5103\times10^9\ m/s$$

$$r_b = \sqrt{\frac{G}{324}\frac{e}{\alpha c^2}} = \frac{1}{18}\sqrt{G}\frac{\sqrt{\alpha\hbar c}}{\alpha c^2} = \frac{1}{18\sqrt{\alpha}}\sqrt{\frac{G\hbar}{c^3}} = \frac{1}{18\sqrt{\alpha}\sqrt{2\pi}}\sqrt{\frac{G\hbar}{c^3}} \tag{A6}$$

$$= 0.259\sqrt{G\,h/c^3}$$

If the "effective" radius of the sphere oscillating at the amplitude of r_b were $1.528r_p$ (instead of $3r_p$) then the ether particle radius would be close to the value of the Planck length.

$$r_b = \sqrt{\frac{G}{324}\frac{e}{\alpha c^2}\frac{m_g}{m_p}} = \sqrt{\frac{.6673\times10^{-10}}{324}\frac{1.5189\times10^{-14}}{0.0072974}\frac{1}{\left(2.9979\times10^{8}\right)^2}} = 1.0510\times10^{-35}\,m$$

$$m_b = 0.05003\frac{G}{e^2}\left(\frac{m_g}{m_p}\right)^2 \alpha^{3/4} m_p^3 = \frac{0.05003\times6.673\times10^{-11}\left(0.0072974\right)^{3/4}\times\left(1.6726\times10^{-27}\right)^3}{\left(1.5189\times10^{-14}\right)^2}$$

$$= 1.6906\times10^{-66}\,kg \tag{A7}$$

$$l = 0.11402\frac{e^2}{\alpha^{5/4}c^2 m_p} = \frac{0.11402\left(1.5189\times10^{-14}\right)^2}{\left(.0072974\right)^{5/4}\left(2.9979\times10^{8}\right)^2\left(1.6726\times10^{-27}\right)}$$

$$= 8.2045\times10^{-17}\,m$$

The four derived constants are

$$\text{v}_m = 3.5103\times10^{9}\,m/s$$

$$r_b = 1.0510\times10^{-35}\,m \tag{A8}$$

$$m_b = 1.6906\times10^{-66}\,kg$$

$$l = 8.2045\times10^{-17}\,m$$

Other parameters often needed are

$$\alpha = 0.0072974$$

$$r_p = \frac{\hbar}{2m_p c} = \frac{1.0546\times10^{-34}}{2\left(1.6726\times10^{-27}\times2.9979\times10^{8}\right)} = 1.0516\times10^{-16}\,m$$

$$\rho_0 = \frac{e^2}{2r_p^{\,4}c^2} = \frac{\left(1.5189\times10^{-14}\right)^2}{2\left(1.0516\times10^{-16}\right)^4\left(2.9979\times10^{8}\right)^2} = 1.0495\times10^{19}\,kg/m^3$$

(A9)

$$\eta = \frac{\rho_0}{m_b} = \frac{1.0495\times10^{19}}{1.6906\times10^{-66}} = 6.2079\times10^{84}\,m^{-3}$$

Appendix B. Further Discussion of the Neutrino Structure

All the analysis in this paper is based on the premise that stable inhomogeneous states can exist in a gas consisting of hard elastic particles which interact only by repulsion when they collide. We are familiar with hurricanes, cyclones, and tornados. These flow patterns are stable inhomogeneous assemblages of air molecules. Comprehensive analyses showing the mechanisms of their stability have not been developed. However, the observed stability of tornados, etc, which are made up of gases with electromagnetic properties and which involve heat flow does not prove that a gas of inert particles could form a stable inhomogeneous state. These flows open the question but do not lead us to a conclusion for or against a stable inhomogeneous state of inert particles.

We also know that all macroscopic inhomogeneous states of inert particles which we observe always dissipate and move toward homogeneity. In contrast, the stable inhomogeneous states we are considering in this paper are contained within a sphere with a diameter of one mean free path of the ether gas, which value is $10^{-16}\,m$. Further, the observed particles of this theory, i.e., the condensed regions of all the inhomogeneous assemblages are in the order of 10^{-23} meters. If the same inhomogeneous phenomenon existed with air molecules as with the basic ether gas particles then the air assemblages would exist inside a sphere with a radius equal the mean free path of $10^{-6}\,m$ and the solid core diameter would be smaller than this. Such assemblages could occur without being recognized as having the structure of the inhomogeneous assemblages of this theory. Possibly such assemblages in the atmosphere could be the origin of hail and raindrops.

Scientists have tried for more than a century to find a stable inhomogeneous state and have tried to prove that a stable inhomogeneous state for kinetic particles can not exist. So far, science has failed in both endeavors. Clearly in any theory of physics there must be some mechanism for producing inhomogeneous states, since the universe is inhomogeneous.

The results of this paper argue strongly for the existence of stable inhomogeneous assemblages of the hypothesized ether gas. Verification of the stable assemblage possibly could be obtained experimentally using gas molecules with a very long mean free path or by a dynamic analysis.

The stable inhomogeneous states which is envisioned for this theory consists of a particle flow configuration which is essentially contained within a sphere with a radius in the order of one mean free path of the background gas. The concept of this theory is to begin with a uniform (homogeneous) gas and construct one basic type of inhomogeneous assembly. This inhomogeneous assembly then forms hydrogen atoms which produce a flow of particles which is the gravitational field. Hydrogen atoms are gravitationally attracted to each other to produce large assemblages (stars) and then other atoms. The basic assemblage of this theory thus not only is inhomogeneous but it also provides a mechanism for producing other inhomogeneties for further organizing the universe.

We have spent an appreciable amount of effort attempting to develop a dynamic analysis of the hypothesized state. We consider the analysis in three different regions:

1. A radial inflow into a spherical sink where the inflow reaches sonic speed at a spherical surface with a diameter in the order of the mean free path of the gas. Initially the sphere is considered to be at rest but later we planned to determine the inflow to the sphere when it translates at the speed of $v_r - v_m$ (the rms speed less the mean speed of the background gas).
2. An inflow from the sonic speed spherical surface using curved stream tubes down to the solid cylindrically shaped core of the assemblage.

3. An axial inflow into the aft end of the solid cylindrical core at velocity v_m, impacts on the sides of the cylindrical surface to completely condense the core and to accelerate it from velocity v_m to velocity v_r, and an output from the front of the core at velocity v_r.

The basic conservation equations governing the particle flows for the envisioned inhomogeneous assemblage are the conservation of mass, linear momentum in three directions, angular momentum in three directions, and energy; eight conservation equations total. In addition there are the equations of state. One usually used is the ideal gas equation $p = \frac{1}{3}\rho v_r^2$; where p is the pressure, ρ is the mass density, and v_r is the rms velocity, which velocities are measured relative to the flowing gas. Another state equation is for the solid in the region where the particles are touching each other. The flow analysis is conventional except for handling the gas equation of state when the gas takes a curved path.

When gas flowing in a straight path becomes perturbed into a curved path half of the particles have thermal velocity components opposite the velocity and half in the same direction as the flow velocity. The half with opposite components have radial centrifugal forces that are smaller than the half with components in the same sense as the flow velocity. As a result there is a transient radial adjustment which accentuates the curvature since the slower particles will move inward and the faster particles will move outward. This transient radial adjustment further accentuates the curvature since removal of the particles with aft thermal components increase the flow velocity at the outer stream tubes. Similarly, adding aft thermal component particles opposite the flow decreases the flow speed at the inner stream tubes. When the steady state is reached and the curvature becomes constant the equation of state may be homogeneous (i.e.,

$p = \frac{1}{3}\rho v_r^2$) or it may be skewed as a result of the inertial field acting upon the particles. In any case the separation of particles due to their thermal velocities (i.e., thermal separation) as a result of curved flow may be the mechanism causing hurricane winds to take their circular paths. Further, tornados, “dust devils”, and turbulent vortices may also be due to this mechanism.

The phenomenon of thermal separation because of a gas taking a curved path is well documented in the case of vortex tubes. A vortex tube is a hollow cylinder into which a gas is injected at the periphery perpendicular to the cylinder longitudinal axis. The air enters and is immediately forced into a spiral path as it rotates and advances along the cylindrical longitudinal axis. At the exit from the cylinder (which is open at least on one end) the gas at the periphery can be tens to a hundred degrees greater than the inlet gas temperature and at the cylinder axis it can be tens to a hundred degrees cooler than the inlet gas.

Let us present some more discussion of tornados and hurricanes to provide additional insight into the mechanism of the stable state envisioned for this theory. Consider first the question of what produces the curved flows in hurricanes and tornados – and, further, for vortices in general. Consider a mass of air moving in a straight path. Assume a low pressure region occurs on one side of the moving mass or that it intercepts another moving mass in such a way that it will produce curvature of the straight path. In order to understand the response of air when deflected into a curved path we need to consider the individual molecule.

Air moving in a straight path has a flow velocity and a thermal velocity. The flow velocity is the average of the components of the individual molecules along the direction for which the average is a maximum. The thermal velocity of a molecule is the molecule velocity less the flow velocity. The thermal speeds at a point have the same distribution along every direction over 4π sterradians. They have the Maxwell-Boltzmann distribution.

When the gas begins to take a curved flow path it is not generally known how the gas behaves. One thing is clear, however, if we divide the molecules into two equal number categories. Half of the molecules have thermal velocities which have components along the flow velocity with the same sense as the flow velocity and half will have components in the opposite sense. The response of the molecules to the "inertial field" of the curved path will depend only upon the magnitude of the individual molecule velocity component parallel to the flow velocity. The average response of that half of the molecules with thermal components adding to the flow velocity, i.e., with an average component of $\text{v}_f + \text{v}_r/2$ where v_f is the flow velocity and v_r is the group thermal velocity, will be a radial inward acceleration of $(\text{v}_f + \text{v}_r/2)/r$ where r is the radius of curvature. The average response of the half of the molecules with thermal components subtracting from the flow velocity will be $(\text{v}_f - \text{v}_r/2)/r$. The forward thermal component molecules will migrate away from the center of curvature and the aft thermal component molecules will migrate toward the center of the curvature. This phenomenon is a thermal separation mechanism.

When the forward thermal component molecules migrate outward the tangential flow velocity at the increased radius will increase because of the different mix of particles. Similarly, when the aft thermal component molecules migrate inward the tangential flow velocity at the decreased radius will decrease as a result of the different mix of particles. We then have the external forces, either a high pressure at the outer radius or a low pressure at inner radius, along with the thermal separation mechanism causing the flow to have curvature. The net result of the thermal separation will be an instability toward increasing curvature of the flow. In a tornado or a hurricane there could be a continual inflow and continual thermal separation so that the "final" state of the air is difficult to predict. However, with a gas in a steady state condition, such as achieved in a drum rotating at constant speed the equilibrium state can be predicted, subject to the assumption that the thermal velocity distribution is homogeneous. We present this analysis now.

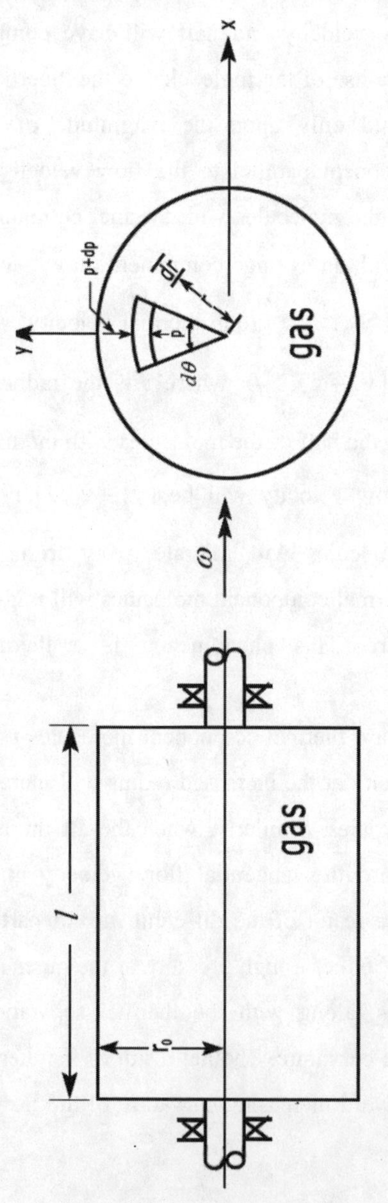

Figure B1 Rotating Drum Filled With Gas

Figure B1 shoes a cylindrical drum filled with a gas and rotating about its cylindrical axis. Summing, forces on the differential element shown we have

$$\rho r d\theta l - (\rho + dp)(r + dr) d\theta l = dm\, A_n / r = -(\rho r d\theta dr l) A_n / r \qquad \text{(B1)}$$

where A_n is the normal acceleration of the "average" gas particle, and is directed inward. A simple way to understand what the square of the average velocity is follows. Figure B2 shows a typical particle with its flow velocity parallel to x and its thermal velocity v_r with components in the x, y, and z directions. The average thermal velocity of all particles with thermal velocity components in positive x direction is $1/2\ v_r$, i.e., the centroid of a hemispherical surface. The average acceleration for both halves of the sphere then is

$$\begin{aligned}(A_n)_{avg} &= \left[\frac{1}{2}\left(v_f + \frac{1}{2}v_r\right)^2 + \frac{1}{2}\left(v_f - \frac{1}{2}v_r\right)^2\right] \Big/ r \\ &= \left(v_f^2 + \frac{1}{4}v_r^2\right) \Big/ r\end{aligned} \qquad \text{(B2)}$$

Combining (B1) and (B2) then simplifying gives

$$rdp = \rho\left(v_f^2 + \frac{1}{4}v_r^2\right)dr - pdr \qquad \text{(B3)}$$

Using $p = (1/3)\rho v_r^2$ we have

$$rdp = \left(3\frac{v_f^2}{v_r^2} - \frac{1}{4}\right)pdr \qquad \text{(B4)}$$

The ratio v_f/v_r is constant since v_f and v_r appear in the dynamic equation in exactly the same form. Further, since v_f varies linearly with the radius (from obvious kinematic requirements) it is necessary for v_r to vary linearly with the radius. The ratio v_f/v_r then must be constant.

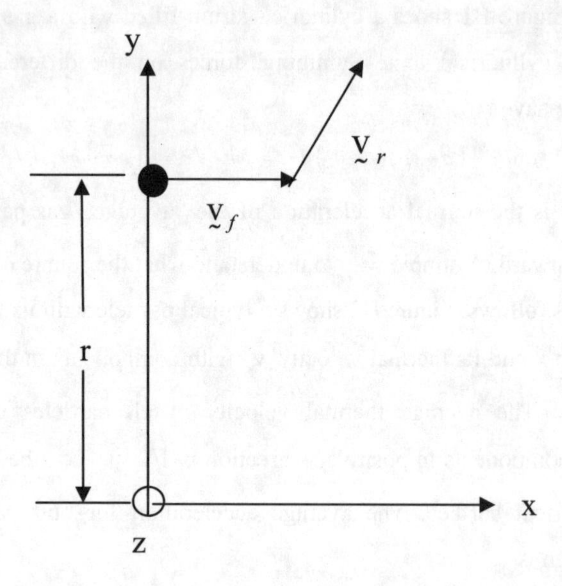

Figure B2. Gas Particle With its Flow and Thermal Velocities

We can now integrate from r_1 to r_0 to give

$$\frac{p}{p_1} = \frac{r^a}{r_1^a} \tag{B5}$$

Where r_1 is an arbitrary location away from the center of rotation, p_1 is the pressure at r_1, and

$$a = 3\frac{\mathrm{v}_f^{\,2}}{\mathrm{v}_r^{\,2}} - \frac{1}{4} \tag{B6}$$

For large values of $\mathrm{v}_f/\mathrm{v}_r$ the pressure increases significantly with radius. The density is determined from the equation

$$\rho = 3p\Big/\mathrm{v}_r^{\,2} = 3p_0\frac{r^a}{r_1^a}\Big/(\omega_t r)^2 \tag{B7}$$

where $\omega_1 r$ is the thermal velocity (ω_t is the thermal velocity constant of proportionality). To have a constant density we set $a-2=0$ or

$$3\text{v}_f/\text{v}_r - 1/4 = 0 \tag{B8}$$

or

$$\text{v}_f/\text{v}_r = \sqrt{3/4} = 0.866 \tag{B9}$$

If $\text{v}_f > 0.866\text{v}_r$ and the path is curved then the density will increase with radius for the steady state rotating drum.

We have shown a possible mechanism for producing curved flow in a gas. Such a flow could be initiated by a random fluctuation of the background producing a small curvature then the curvature gets accentuated by the thermal separation process described here. If such a flow can be made to exceed the nozzle critical speed which is $\left(\sqrt{5/12}\right)\text{v}_r = 0.645\text{v}_r$ [1] then the curved flow could cause a density increase and thus a condensation of the particles. The thermal separation and the condensation phenomena are required in the stable state envisioned for this theory. Let us now investigate flows into a sink then out of the sink. Consider a sphere with two symmetric holes in it through which air can enter the sphere. Let a vacuum pump be placed at the bottom of the sphere, see Figure B3. Now when air is pumped from the sphere air beings flowing in through the two inlet holes. "Flow smoke" can be introduced separately into each of the holes to visualize the flow for each hole.

It is expected that air will begin swirling about the y-axis clockwise when viewed looking inward toward the +y axis, because of the flow curvature required for holes not on the y-axis, and the resulting angular momentum produced. Figure B3 shows the flow for two holes. It is seen

[1] The speed of sound "a" is $\sqrt{(c_p/(3c_v))}\,\text{v}_r$, where c_p is the specific heat at constant pressure and c_v at constant volume. For air c_p/c_v=1.4 so $a=\sqrt{1.4/3}\text{v}_r$. For an ideal gas c_p/c_v=5/3 and $a = \sqrt{5/(3\times3)}\text{v}_r = 0.745\text{v}_r$.

that angular momentum generated will produce rotation in the –y direction. This figure clearly shows the generation of the angular momentum. However, this experiment will have to be extended greatly beyond this to prove the existence of the structure of the neutrino. We now present a simple analysis of the flow into a spherical sink anticipated for the neutrino.

The boundary condition for the radial inflow is assumed to be such as to have back pressure less than the critical pressure at the mean free path diameter sphere. Basically it is assumed that the pressure is low inside the sphere and the flow into the sphere is from an infinite volume background whose characteristics many diameters distant from the sphere are essentially those of the undisturbed background ether gas. Just inside the sphere the gas

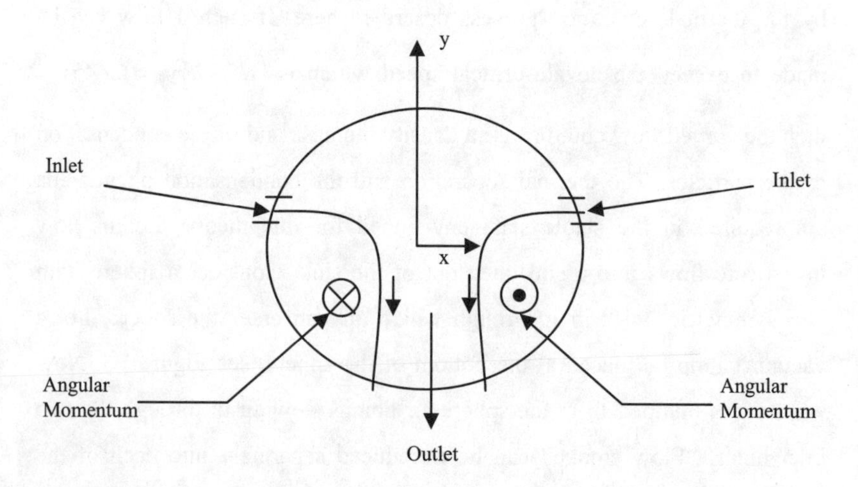

Figure B3. Air Flow Showing the Angular Momentum Produced

flows inward along curved stream tubes, becomes completely condensed, then comes out of the sphere as a very small cross-section flow ($10^{-25}\,m$ diameter) at velocity v_r . The area of the output flowing mass is $\left(10^{-25}/10^{-16}\right)^2 = 10^{-18}$ times the area of the input and thus the output area is negligible in comparison with the input area. The inflow is essentially spherically symmetric even through there is a provision for the mass flow out to be equal the inflow. This portion of the flow making up the hypothesized inhomogeneous assemblage is straightforward and its analysis is simple.

A fairly accurate approximation to the inflow is to assume that the density remains constant down to the mean-free path sphere. Further, we take the flow velocity to be the sonic velocity which is approximately $0.8\mathrm{v}_m \left(\doteq 3\times10^9\ m/s\right)$. Now the input flow $\dot{m}$ is

$$\begin{aligned} \dot{m} &\doteq \rho A\mathrm{v} = m_b\eta\left(4\pi\right)l^2\times0.8\mathrm{v}_m \\ &\doteq 10^{-66}\times10^{85}\left(4\pi\right)4\times10^{-32}\times3\times10^9 \qquad \text{(B10)} \\ &\doteq 10^{-2}\,kg/s \end{aligned}$$

A more accurate calculation (accounting for compressibility) can be performed using a lumped-parameter analysis which consists of writing the radial (linear) momentum equation, the energy balance, and the mass balance with the equation of state $p_0 = 1/3\,\rho\mathrm{v}_r^{\,2}$. In the following paragraphs we present the compressible analysis.

The flow of gas into a three dimensional "sink" is the same as the flow down a conical converging nozzle where the cone half angle is increased to 180°. The flow of a compressible gas along a converging nozzle is straightforward. Let us now present the analysis for the flow along a straight converging nozzle.

The converging nozzle gas flow analysis is presented on pages 298 – 300 of Binder [16]. Considering the case of flow from an ambient source, when the flow velocity is zero the velocity at the exit is

$$\mathrm{v}_2^{\,2} = \frac{2k}{k-1}\frac{p_1}{\rho_1}\left[1-\left(\frac{p_2}{p_1}\right)^{\frac{k-1}{k}}\right] \qquad \text{(B11)}$$

where k is the specific heat ratio c_p/c_v which for air is 1.4 and for the hard particle gas of this theory is $5/3$, p_1 and ρ_1 are the ambient gas pressure and mass density, and p_2 is the exit pressure. As p_2 decreases it reaches a value which maximizes v_2. This magnitude of pressure is the "critical" pressure and its value is

$$p_c = p_1 \left(\frac{2}{k+1} \right)^{\frac{k}{k-1}} \tag{B12}$$

Substituting this as p_2 in (B11) gives the maximum outflow velocity v_c as

$$v_c^{\ 2} = \frac{2k}{k-1} \frac{p_1}{\rho_1} \left[1 - \frac{2}{k+1} \right] = \frac{2k}{k+1} \frac{p_1}{\rho_1} \tag{B13}$$

For the ideal gas with $k = 5/3$ and using $p_1 = \rho_1 v_{r1}/3$, where v_{r1} is the ambient thermal velocity. The velocity becomes

$$v_c = \sqrt{5/12} v_{r1} = 0.645 v_{r1} \tag{B14}$$

we note that this maximum velocity is smaller than the speed of sound $(a = 0.745 v_r)$ since

$$\frac{v_c}{a} = \frac{\sqrt{5/12} v_{r1}}{\sqrt{5/9} v_{r1}} = 0.866 \tag{B15}$$

The density at the nozzle outlet is computed now.

$$\frac{p_1}{\rho_1^{\ k}} = \frac{p_c}{\rho_c^{\ k}} = \frac{p_c}{\rho_1^{\ k}} \left(\frac{2}{k+1} \right)^{\frac{k}{k-1}} \tag{B16}$$

Thus

$$\rho_c^{\ k} = \rho_1^{\ k} \left(\frac{2}{k+1} \right)^{\frac{k}{k-1}} \tag{B17}$$

or

$$\rho_c = \rho_1 \left(\frac{2}{k+1} \right)^{\frac{1}{k-1}} \tag{B18}$$

For the ideal gas

$$\rho_c = 0.650 \rho_1 \tag{B19}$$

The critical density for air (with k = 1.4) is

$$\rho_c = 0.634\rho_1 \tag{B20}$$

The requirement that the angular momentum $(\hbar/2)$ for these assemblages to be always the same implies that the mass inflow rate always to be one precise value independent of what occurs inside the radius at which sonic speed occurs. It is clear that this radius must be in the order of the mean free path. Once such a flow were initiated the mass flow probably would increase until it is impeded. The impediment would be produced by background particles from the opposite end of each diameter interfering with inflow at the other end of the diameter.

It is commonly believed that further inflow beyond the critical sphere is impossible, since such flow is still converging. We have presented arguments which might provide a mechanism for further converging flow. These arguments are based on the flow taking a curved path which results in higher (total) speed particles moving toward a larger radius path. The curved paths may provide thermal separation and thus may permit further converging flow. For the time being we will assume that further flow is possible and that it results in complete condensation.

First we assume a spherically symmetric inflow and then determine the diameter of the sphere at which complete condensation occurs. The mass inflow at the sonic radius is

$$\dot{m} = \rho A \mathrm{v} = m_b \eta (4\pi) l^2 \times 0.8 \mathrm{v}_m \tag{B21}$$

and at the completely condensed radius r_v is

$$\dot{m} = \rho_{solid}\left(4\pi r_v^{\ 2}\right) \times \mathrm{v}_m \tag{B22}$$

where we assume that the particles are flowing in at the background mean speed v_m. Now

$$\rho_{solid} = \frac{m_b}{\left(4/3\pi r_b^{\ 3}\right)} \tag{B23}$$

Using ρ from above and equating $\dot{m}$ from (B21) to (B22) gives

$$r_v \doteq \left[4/3\pi(0.8)\eta r_b^3\right]^{1/2} l \doteq 3\times10^{-26}\,m. \qquad \text{(B24)}$$

The next step in the analysis, which we have not taken, is let the assemblage move at the velocity $v_r - v_m$ and then try to fit stream-like into the spherical volume subject to the following conditions:

1. The tubes fill the volume down to a pre-selected cylindrical core.
2. The tubes and radius are selected so that no friction results between the tubes.
3. The flow produces an angular momentum of $\hbar/2$ about the core cylindrical axis.
4. The flow is all turned so that it is directed axially along the cylindrical axis in the direction of the cylindrical core translation (and at velocity $v_r - v_m$).
5. The flow also must apply a thrust to the side of the cylindrical core equal to the proton thrust ($\doteq 10^6$ Newtons).

The last step in the analysis is to determine the size of the cylindrical solid core. Hopefully some additional insight into the core geometry will come from the above analysis. (We have made approximations of its size if it were spherical.) The flow into the aft end of the core would be at velocity v_m. The axial thrust on the side would increase this velocity to v_r as the final condensation takes place. The mass and energy conservation equations for the core are straight forward. Assume that all the particles entering the core are a sample of the background gas and that they are all aligned parallel to each other, but not touching each other. Let the particles be accelerated from velocity v_m to v_r by the side thrusting force while the final condensation takes place. The mass flow rate in is $\dot{m}$, at velocity v_m, the mass flow rate out is $\dot{m}$ at v_r, and the mass continuity equation is $(\rho A)_{in}\ v_m = (\rho A)_{out}\ v_r$. The energy balance is

$$(1/2)(\rho A)_{in}\ v_m v_r^2 = (1/2)(\rho A)_{out}\ v_r v_r^2 \qquad \text{(B25)}$$

which is seen to be balanced by substituting from the mass continuity equation. Figure B4 shows the core flows. The momentum balance equation is

$$\left[(\rho A)_0 \mathrm{v}_r\right]\mathrm{v}_r = \left[(\rho A)_i \mathrm{v}_m\right]\mathrm{v}_m + F \tag{B26}$$

where F is the total axial force on the core. Using mass continuity we can write (B26) as

$$\mathrm{v}_r - \mathrm{v}_m = F / \left[(\rho A)_i \mathrm{v}_m\right] \tag{B27}$$

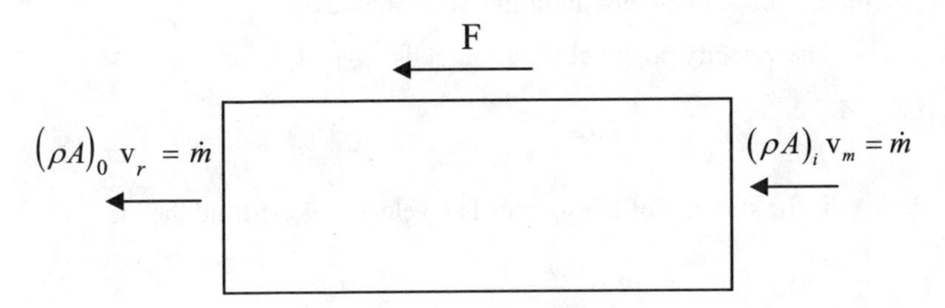

Figure B4. Core Flows

the force F propels the core at the speed of light $(c = \mathrm{v}_r - \mathrm{v}_m)$. The other half of the force pair acts on the background. The cores presumably can exist with a wide range of masses.

The assemblage we have just described may be an *organizing* mechanism increasing the flow velocity by 8% which is the opposite of the *disorganizing* mechanism described earlier in this book where energy was held constant and the mean speed went from v_r to v_m, an 8% drop in velocity.

Appendix C. Inverse Square Forces in a Hydrodynamic Medium

Two solids immersed in a fluid (liquid or gas) and separated by a distance large in comparison to the size of the solids can attract or repel each other by certain types of small motions or deformations of the solids. Whittaker [6] discuss this for a number of different cases. In the following paragraphs we develop theoretical analysis for the case of a pair of breathing spheres immersed in a hydrodynamic medium. A hydrodynamic medium is a non-viscid incompressible fluid. The analysis follows that presented by Basset [10].

1. Sources, Sinks, Doublets, and Line Sink-Sources

The velocity potential for a source is given by

$$\varphi = -\frac{m}{r} \tag{C1}$$

where m is the strength of the source. The velocity of the fluid then is

$$\mathrm{v} = \frac{d\varphi}{dr} = \frac{m}{r^2} \tag{C2}$$

The velocity is directed outward from the source homogeneously over all directions. The quantity of flow Q, is

$$Q = A\mathrm{v} = 4\pi r^2\left(\frac{m}{r^2}\right) = 4\pi m \tag{C3}$$

and the units are volume per unit time. For a sink the strength is taken as *–m*. The flow is inward and is given by (C2) and (C3).

A doublet is formed by an adjacent sink and source pair which, in the limit, are at the same point. We compute the velocity at point "a" which is at a distance r from the origin "o" and at the angle α in Figure C1.

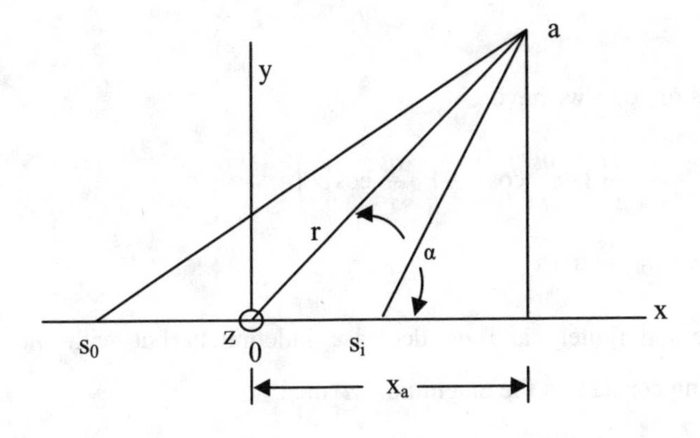

Figure C1. Doublet

Due to the sink at s_i, the velocity potential is $\varphi_i = -m/(s_i a)$ and due to the source at s_o it is $\varphi_0 = m/(s_0 a)$. Thus, due to the source and sink the potential is

$$\varphi = \varphi_i + \varphi_0 = \frac{-m}{s_i a} + \frac{m}{s_0 a} \tag{C4}$$

now

$$os_i / r = \cos\alpha \tag{C5}$$

and, let

$$os_0 = os_i = os \tag{C6}$$

we have

$$\begin{aligned} s_i a &= r - os_i \cos\alpha \\ s_o a &= r + os_i \cos\alpha \end{aligned} \tag{C7}$$

Thus

$$\begin{aligned} \varphi &= -\frac{m}{r - os\cos\alpha} + \frac{m}{r + os\cos\alpha} \\ &= \frac{m}{r}\left(\frac{1}{1 - \frac{os}{r}\cos\alpha} - \frac{1}{1 + \frac{os}{r}\cos\alpha} \right) \end{aligned} \tag{C8}$$

For small values of os/r we have

$$\begin{aligned}\varphi &= \frac{m}{r}\left(1+\frac{os}{r}\cos\alpha-1-\frac{os}{r}\cos\alpha\right)\\ &= \frac{os}{r^2}\cos\alpha\end{aligned} \tag{C9}$$

Let m increase indefinitely and os decrease indefinitely but with the product remaining constant at the magnitude μ then

$$\varphi = \frac{\mu}{r^2}\cos\alpha = \frac{\mu x_a}{r^3} \tag{C10}$$

where x_a is the x coordinate of the point "a".

The velocity potential due to a sheet of doublets with strength "m" per unit of surface area where the doublet axis is in the direction of the normal to the sheet is given by

$$\varphi = -\iint \frac{m\cos\varepsilon\mu}{r^2}ds \tag{C11}$$

where ε is the angle from the normal. Integrating (C11) gives

$$\varphi = -\iint m d\Omega \tag{C12}$$

where Ω is the solid angle (in sterradians). For a constant value of m

$$\varphi = -m\Omega \tag{C13}$$

When the area is a narrow long rectangle along the z-axis Laplace's equation (of continuity) becomes

$$\frac{d^2\varphi}{dr^2}+\frac{1}{r}\frac{d\varphi}{dr}=0 \tag{C14}$$

The solution of (C14) is

$$\varphi = m\ln r \tag{C15}$$

and

$$\frac{d\varphi}{dr}=\frac{m}{r}=\text{v} \tag{C16}$$

The φ given by (C15) is a line source of infinite length whose strength per unit length is "m". In this case the flow appears as shown in figure C2.

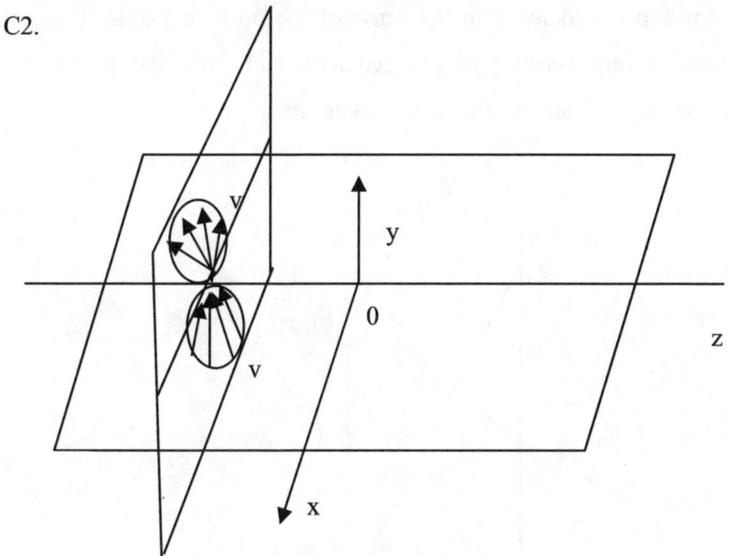

Figure C2. Flow for a Line Source

The flow is invariant with z.

2. Images

In developing the analysis we will use the concept of an image. A hydrodynamic system H is a system consisting of an inviscid and incompressible fluid flowing in a prescribed volume. Let H_1 and H_2 be two hydrodynamic systems. The streamlines either form closed curves or have their extremities in singular points (such as a source or sink)

Let s and s' be points at two sources whose strengths are m, see Figure C3. Let A bisect the distance between s and s' and construct the perpendicular plane ab bisecting the lines ss' . The component of velocity normal to this plane is

$$+\frac{m}{sb^2}\cos bsa - \frac{m}{sb^2}\cos bs'a = 0 \tag{C17}$$

where a positive velocity is in the sense of the positive x-axis. The component is zero because of the geometry and since the source strengths are equal. Thus the flux across plane ab is zero.

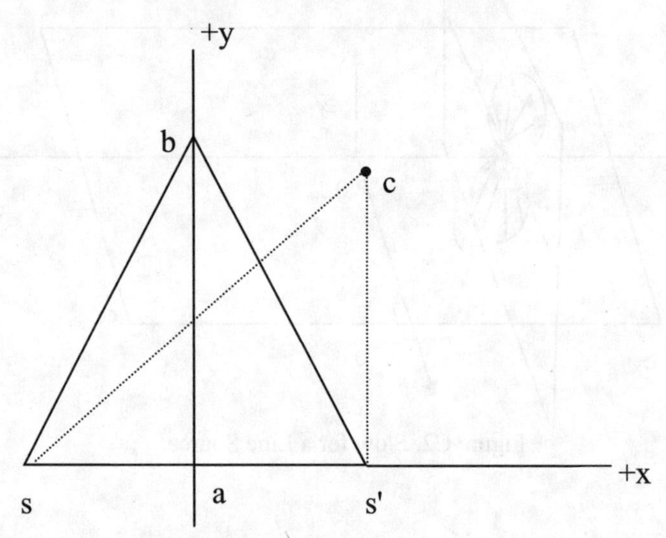

Figure C3. Two Sources

Let us now determine the velocity potential at any point "c" to the right of plane ab for a source at s' where there is an impregnable plane at ab. What we do is ignore the plane and place an equal strength source at the point s, where $sa = s'a$. Now, the velocity potential at "c" due only to the source at s' (and an impregnable plane ab) is

$$\phi = -\frac{m}{cs'} - \frac{m}{cs} \tag{C18}$$

The image of the source at s' with respect to a plane is an equal strength source situated at a point s on the other side of the plane whose distance from it is equal to that of s'.

Now let us find the image of a source placed outside a sphere. The spherical surface in this case is the impregnable surface. Figure C4 shows a sphere of radius "*a*", a source at *s*, a point *Q* on the sphere, and a point *P* outside the sphere at a distance *r* from the sphere center.

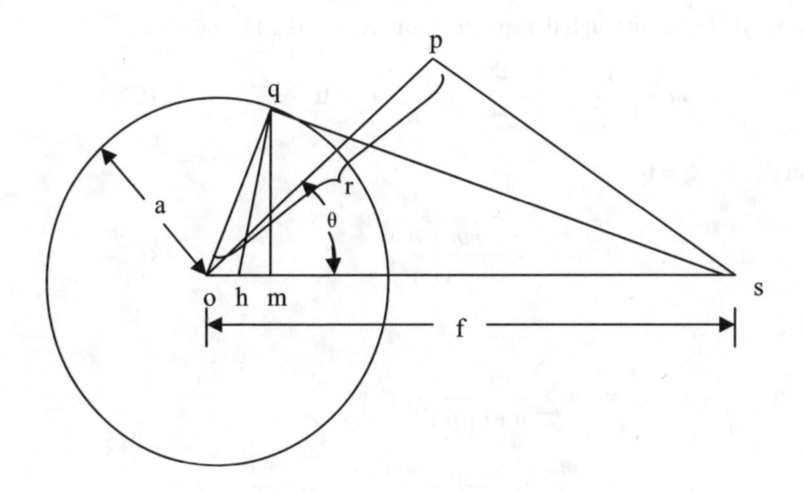

Figure C4. Sphere and Outside Source

The velocity potential at *P* due to the source of strength *m* at *s* is

$$\varphi = -\frac{m}{sP} = -\frac{m}{\left(r^2 - 2fr\mu + f^2\right)^{1/2}} \tag{C19}$$

where $f = os$ and $\cos\theta = \mu$. For all points close to the sphere (where $r < f$) φ can be expanded in the series

$$\varphi = -\frac{m}{f} - \frac{m}{f}\sum_{n=1}^{n=\infty}\left(\frac{r}{f}\right)^n P_n(\mu) \tag{C20}$$

where $P_n(\mu)$ is the zonal harmonic of degree n.

At all points outside the sphere the velocity φ' for the image of s can be expanded in the series

$$\varphi' = \frac{1}{r}\sum_{n=1}^{n=\infty} A_n \left(\frac{a}{r}\right)^n P_n \tag{C21}$$

With the sphere at rest for $a = r$

$$\frac{d\varphi}{dt} + \frac{d\varphi'}{dt} = 0 \tag{C22}$$

since no flux goes through the sphere. From (C20), (C21), and (C22)

$$m\sum_{n=1}^{n=\infty} \frac{na^{n-1}}{f^{n+1}} P_n + \sum_{n=1}^{n=\infty} A_n \frac{n+1}{R^2} P_n = 0 \tag{C23}$$

From (C23) $A_0 = 0$ and

$$A_n = -\frac{mn}{n+1}\left(\frac{a}{f}\right)^{n+1} \tag{C24}$$

and

$$\begin{aligned} \varphi' &= -m\sum_{n=1}^{n=\infty} \frac{n}{n+1} \frac{a^{2n+1}}{(fr)^{n+1}} P_n \\ &= -\frac{ma}{f}\sum_{n=1}^{n=\infty} \frac{c^n}{r^{n+1}} P_n + \frac{ma}{f}\sum_{n=1}^{n=\infty} \frac{c^n}{r^{n+1}} \frac{P_n}{n+1} \end{aligned} \tag{C25}$$

where

$$c = a^2/f \tag{C26}$$

For $c < r$ (or $a^2 < fr$) and using the dummy variable $x < c$ we have

$$\int_0^c \frac{d\lambda}{\left(r^2 - 2\lambda r\mu + \lambda^2\right)} = \sum_{n=0}^{n=\infty} \left(\frac{c}{r}\right)^{n+1} \frac{P_n}{n+1} \tag{C27}$$

by adding and subtracting $ma/(fr)$ from (C25) φ' is

$$\varphi' = -\frac{ma}{f} \frac{1}{\left(r^2 - 2rc\mu + c^2\right)^{1/2}} + \frac{ma}{f} \frac{1}{\left(r^2 - 2r\lambda\mu + \lambda^2\right)^{1/2}} \tag{C28}$$

The first term is a point source of strength ma/f located at H where $OH = c = a^2/f$ and the second term is a line sink with strength m/a per unit of length. The sink extends from H to O.

We now have the remarkable result that $\varphi + \varphi'$ gives the velocity potential for a fixed source outside an impregnable sphere fixed in space and immersed in a hydrodynamic medium.

As a preliminary to the breathing sphere analysis we need to prove the additional following theorem. In Figure C5 let $PM = \omega,\ AM = Z,\quad BM = Z',\quad AB = c,\ \cos\theta = \mu,$ and $\cos\theta' = \mu'$.

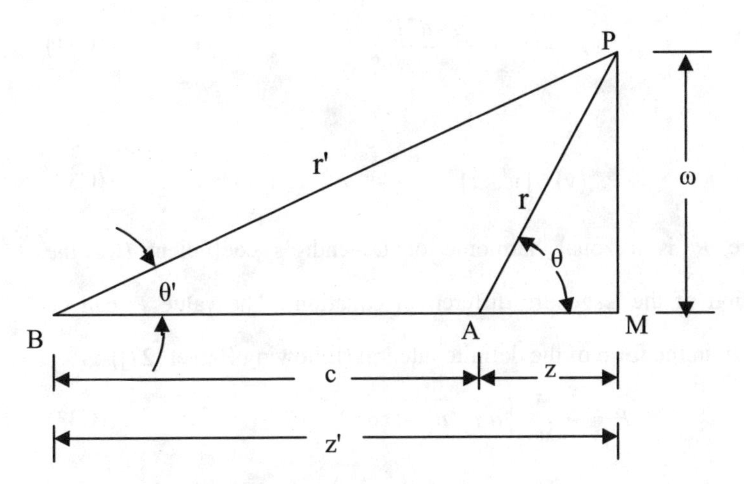

Figure C5. Two Right Triangle With a Common Altitude

Let $P_n^m(\mu)$ be an associated function of degree n and order m whose origin is at A and whose axis is AM. Let $P'^m_n(\mu')$ be a similar function whose origin is at B. For $r < c$ we want to prove

$$\frac{P'^m_n}{r'^{\,n+1}} = \frac{r^m}{(n-m)!c^{n+m+1}}\left[\frac{(n+m)!}{2m!}P_n^m - \frac{(n+m+1)!r}{(2m+1)!c}P_{m+1}^m + \dots \right.$$
$$\left. + \frac{(-)^s(n+m+s)!}{(2m+s)!}\left(\frac{r}{c}\right)^s P_{m+n}^m + \dots\right] \qquad \text{(C29)}$$

and for $r' < c$

$$\frac{(-)^{n-m} P_n^m}{r^{n+1}} = \frac{r'^m}{(n-m)!c^{n+m+1}} \left[\frac{(n+m)!}{2m!} P_n^m + \frac{(n+m+1)!}{(2m+1)!} \frac{r'}{c} P'^m_{m+1} + \right. \tag{C30}$$
$$\left. + \frac{(n+m+s)!}{(2m+s)!} \left(\frac{r'}{c} \right)^s P'^m_{m+n} + \ldots \right]$$

To begin this proof we start with the equations (taken from Jahnke [25])

$$P_n^m(\mu) = \left(1-\mu^2\right)^{\mu/2} \frac{d^m P_n(\mu)}{d\mu^m} \qquad \mu < 1 \tag{C31}$$

and

$$P_m^n(v) = \left(v^2 - 1\right)^{m/2} \frac{d^m P_n}{dv^m} \qquad v > 1 \tag{C32}$$

where P_n is a zonal harmonic or Legendre's coefficient (for the solution of the Legendre differential equation). The value P_n can be written in the form of the definite integral (following Basset [27])

$$P_n = \frac{1}{\pi} \int_0^{\pi} \left\{ \mu + \sqrt{\mu^2 - 1} \cos\theta \right\}^m d\theta \tag{C33}$$

or, as

$$P_n = \frac{1}{\pi} \int_0^{\pi} \left\{ \mu + \sqrt{\mu^2 - 1} \cos\theta \right\}^{-n-1} d\theta \tag{C34}$$

An expression for P_n^m will be obtained now. Let

$$V_m = \int_0^{\pi} \frac{\sin^{2m}\theta d\theta}{\left[\mu + \sqrt{\mu^2 - 1} \cos\theta\right]^{n+m+1}} \tag{C35}$$

Differentiating with respect to μ

$$\frac{dV_m}{d\mu}=-\frac{n+m+1}{\sqrt{\mu^2-1}}\int_0^\pi \frac{\left\{\sqrt{\mu^2-1}+\mu+\cos\theta\right\}\sin^{2m}\theta d\theta}{\left\{\mu+\sqrt{\mu^2-1}\cos\theta\right\}^{n+m+2}}$$

$$=-\frac{n+m+1}{\sqrt{\mu^2-1}}\int_0^\pi \frac{\cos\theta\sin^{2m}\theta d\theta}{\left\{\mu+\sqrt{\mu^2-1}\cos\theta\right\}^{n+m+1}}-(n+m+1)V_{m+1} \tag{C36}$$

Integrating by parts

$$\frac{dV_m}{d\mu}=-\frac{(n+m+1)(n-m)}{2m+1}V_{m+1} \tag{C37}$$

Since $V_0=\pi P_m$ we have

$$\frac{d^m P_n}{d\mu^m}=-\frac{(n-m+1)(n-m+2)\ \cdots\ (n-m)}{1\bullet 3\bullet 5\cdots(2m-1)^\pi}V_m \tag{C38}$$

From this

$$P_n^m=\frac{(n+m)!\left(1-\mu^2\right)^{m/2}}{\pi(n-m)!1\bullet 3\bullet 5\cdots(2m-1)}\int_0^\pi \frac{\sin^{2m}\theta d\theta}{\left\{\mu+\sqrt{\mu^2-1}\cos\theta\right\}^{n+m+1}} \tag{C39}$$

Using the transformation

$$\cos\theta=\frac{\mu\cos\varphi+\sqrt{\mu^2-1}}{\mu+\sqrt{\mu^2-1}\cos\varphi} \tag{C40}$$

We get

$$P_n^m=\frac{(n+m)!\left(1-\mu^2\right)^{m/2}}{\pi(n-m)!1\bullet 3\bullet 5\cdots(2m-1)}\int_0^\pi \left\{\mu-\sqrt{\mu^2-1}\cos\varphi\right\}^{n-m}\sin^{2m}\varphi d\varphi \tag{C41}$$

We have just proved that P_n^m can be expressed in either of the two forms, (C41)

$$P_n^m=m\left(1-\mu^2\right)^{m/2}\int_0^\pi\left[\mu-\sqrt{\mu^2-1}\cos\varphi\right]^{n-m}\sin^{2m}\varphi d\varphi \tag{C42}$$

or, (C39)

$$P_n^m = m\left(1-\mu^2\right)^{m/2}\int_0^\pi \frac{\sin^{2m}\varphi d\varphi}{\left\{\mu+\sqrt{\mu^2-1}\cos\varphi\right\}^{n+m+1}} \tag{C43}$$

where

$$M = \frac{(n+m)!}{(n-m)!1\bullet 3\cdots(2m-1)\pi} \tag{C44}$$

Using this we can write (C42) as

$$\begin{aligned}\frac{P_n'^{\,m}(\mu')}{r'^{\,n+1}} &= m\omega^m\int_0^\pi \frac{\sin^{2m}\varphi d\varphi}{(z'+i\varpi\cos\varphi)^{n+m+2}} \\ &= m\omega^m\int_0^\pi \frac{\sin^{2m}\varphi d\varphi}{\left(c+r\left\{\mu+\sqrt{\mu^2-1}\cos\varphi\right\}\right)^{n+m+2}}\end{aligned} \tag{C45}$$

We let

$$\lambda = \mu+\sqrt{\mu^2-1}\cos\varphi \tag{C46}$$

then, for $r < c$

$$\begin{aligned}\frac{P_n'^{\,m}(\mu')}{r'^{\,n+1}} = \frac{m\omega^m}{c^{n+m+1}}\int_0^\pi &\left\{1-(n+m+1)\frac{r\lambda}{c}+\frac{(n+m+1)(n+m+2)}{2!}\left(\frac{r\lambda}{c}\right)^2+\right. \\ &\left.\cdots+\frac{(-)^s(n+m+1)\cdots(n+m+s)}{s!}\left(\frac{r\lambda}{c}\right)+\cdots\right\}\sin^{2m}\varphi d\varphi\end{aligned} \tag{C47}$$

Using the first form of P_n^m, (C42) we find

$$\begin{aligned}\frac{P_n'^{\,m}(\mu')}{r'^{\,n+1}} = \frac{r^m}{(n-m)!c^{n+m+1}}&\left[\frac{(n+m)!}{2m!}P_m^{\,m}-\frac{(n+m+1)!}{(2m+1)!}\frac{r}{c}P_{m+1}^{\,m}+\right. \\ &\left.\cdots+\frac{(-)^s(n+m+s)!}{(2m+s)!}\left(\frac{r}{c}\right)^s P_{m+s}^{\,m}\right\}\end{aligned} \tag{C48}$$

Further, for $r' < c$ we change θ and θ' to their supplements and since

$$P_n^m\left\{\cos(\pi-\theta)\right\} = (-)^{n-m}P_n^m\cos\theta \tag{C49}$$

we obtain

$$\frac{(-1)^{n-m} P_n^m(\mu)}{r^{n+1}} = \frac{r'^m}{(n-m)!c^{n+m+1}}\left[\frac{(n+m)!}{2m!}P'^m_m + \frac{(n+m+1)!}{(2m+1)!}\frac{r'}{c}P'^{\ m}_{m+1} + \cdots + \frac{(n+m+s)!}{(2m+s)!}\left(\frac{r'}{c}\right)^s P'^m_{m+s}\right. \tag{C50}$$

3. The Force of Interaction of Two Breathing Spheres

Let us now consider two spheres A and B immersed in a hydrodynamic medium. The spheres are configured so that they can expand and contract sinuosidally, but always remains spherical. Such spheres are called breathing spheres. Our object is to determine the force of interaction between two such spheres.

Let φ_1 be the velocity potential of the medium where sphere A pulsates and where sphere B doesn't. Similarly let φ_2 be the velocity potential when B pulsates and A doesn't. The velocity potential when both pulsate is

$$\varphi = \varphi_1 + \varphi_2 \tag{C51}$$

The radius of A is a and B is b. If B were not present the value of φ_1 would be

$$\varphi_1 = -a^2\,\dot{a}/r \tag{C52}$$

which can be seen from differentiating φ_1. Thus

$$\mathrm{v}_1 = \frac{a^2\dot{a}}{r^2} \tag{C53}$$

Where $r = a$, $\mathrm{v}_1 = a$, as required, and the velocity is an inverse square value as required. This is the velocity due to a source of strength $a^2\dot{a}$ located at the center of A.

The image of this flow in B will be a source of strength $a^2 b\dot{a}/c$ at the inverse point P coupled with a line sink extending from the inverse point P to the center B. Its strength is $a^2\,\dot{a}/b$ per unit length. Using

$$\begin{aligned} m &= a^2 b\,\dot{a}/c \\ f &= b^2/c \end{aligned} \tag{C54}$$

the strength of the source at P is m and of the line sink from B to P is $-m/f$ per unit length. The image of these in A is given similarly, see (C50). Thus φ_1 and φ_2 will be the potentials of two infinite systems of sources and line sinks which lie within each sphere, respectively.

Let F_2 be the resultant force of the pressure of the liquid on B toward A. Now

$$\begin{aligned} F_2 &= -\iint p\cos\theta ds \\ &= \pi\rho b^2 \int_0^\pi \left(\varphi' + \frac{1}{2}V^2\right)\sin 2\theta d\theta \end{aligned} \tag{C55}$$

where ρ is the density of the medium, b is the radius of sphere B, and V is the velocity of the fluid at the surface of B ($1/2\,\rho V^2$ is the dynamic pressure). Let

$$\begin{aligned} P &= \int_0^\pi \varphi \sin 2\theta d\theta \\ Q &= \int_0^\pi \varphi^2 \sin 2\theta d\theta \end{aligned} \tag{C56}$$

then

$$\int_0^\pi \dot{\varphi} \sin 2\theta d\theta = \dot{P} \tag{C57}$$

In order to find the portion of F_2 which depends upon V^2 let $\mathrm{v} = b^2\dot{b}$ then

$$V^2 = \mathrm{v}^2/b^4 + \left(1/b^2\right)\left(d\varphi/d\theta\right)^2 \tag{C58}$$

Since v is constant over the surface of B (due to our assumed breathing sphere configuration) the portion of F_2 depending upon it is zero. Letting the portion of the pressure depending upon V^2 be I we have

$$\begin{aligned} I &= \frac{1}{2}\pi\rho b^2 \int_0^\pi \mathrm{v}^2 \sin 2\theta d\theta = \frac{1}{2}\pi\rho \int_0^\pi \left(\frac{d\varphi}{d\theta}\right)^2 \sin 2\theta d\theta \\ &= -\frac{1}{2}\pi\rho \int_0^\pi \left(\varphi\frac{d^2\varphi}{d\theta^2}\sin 2\theta + \frac{d\varphi^2}{d\theta}\cos 2\theta\right)d\theta \\ &= -\frac{1}{2}\pi\rho \int_0^\pi \varphi\left(\frac{d^2\varphi}{d\theta^2} + 2\varphi\right)\sin 2\theta d\theta - \frac{1}{2}\pi\rho\left[\varphi^2\right]_0^\pi \end{aligned} \tag{C59}$$

From the Laplace continuity equation

$$-\frac{d^2\varphi}{d\theta^2} = 2b\frac{d\varphi}{dr} + b^2\frac{d^2\varphi}{dr^2} + \cot\frac{d\varphi}{d\theta} \tag{C60}$$

In this equation we set $r = b$ after differentiating. Thus

$$\frac{d\varphi}{dr} = \frac{\mathrm{v}}{b^2} \tag{C61}$$

and

$$I = \frac{1}{2}\pi\rho \int_0^\pi \varphi\left\{\left(\frac{2\mathrm{v}}{b} + b^2\frac{d^2\varphi}{dr^2} - 2\varphi\right)\sin 2\theta + 2\cos^2\theta\frac{d\varphi}{d\theta} - \frac{1}{2}\pi\rho\left[\varphi^2\right]_0^\pi\right\} \tag{C62}$$

Also

$$2\int_0^\pi \cos^2\theta\varphi\frac{d\varphi}{d\theta}d\theta = \int_0^\pi \varphi^2\sin 2\theta + \left[\varphi^2\right]_0^\pi = Q + \left[\varphi^2\right]_0^\pi \tag{C63}$$

And

$$\int_0^\pi \varphi\frac{d^2\varphi}{dr^2}\sin 2\theta d\theta = \int_0^\pi \left(\frac{1}{2}\frac{d^2\varphi^2}{dr^2} - \frac{\mathrm{v}^2}{b^2}\right)\sin 2\theta d\theta = \frac{1}{2}\frac{d^2Q}{dr^2} \tag{C64}$$

From this

$$I = \frac{1}{2}\pi\rho\left(\frac{2\mathrm{v}}{b}\rho - Q + \frac{1}{2}b^2\frac{d^2Q}{dr^2}\right) \tag{C65}$$

And, when $r = b$

$$F_2 = \pi\rho\left(b^2\dot{P} + \frac{v}{b}P - \frac{1}{2}Q + \frac{1}{4}b^2\frac{d^2Q}{dr^2}\right) \tag{C66}$$

Let P_1 be the part of P due to φ_1 then if μ_n is the strength of any image whose distance from B is r, the portion of P_1 due to this is $-2\int_0^{\pi}(\mu\sin\theta\cos\theta d\theta)/\left(b^2+r^2-2br\cos\theta\right)^{1/2}$ and this integral is $-4\mu_n b/\left(3r^2\right)$ if $r > b$ and $-4\mu_n r/\left(3b^2\right)$ if $r < b$.

If μ_m is the strength of the nth source image in A from A, and ρ_n' that of the other extremity of the line sink image, the part of P_1 due to μ_n is

$$\begin{aligned} X_n &= -\frac{4\mu_n b}{3(c-\rho_n)^2} + \frac{4}{3}\int_{\rho'}^{\rho_n}\frac{\mu_n b dx}{(\rho_n-\rho_n')(c-x)^2} \\ &= -\frac{4\mu_n b(\rho_n-\rho_n')}{3(c-\rho_n)^2(c-\rho_n')} \end{aligned} \tag{C67}$$

Let v_n be the strength of the nth image in B, σ_n, σ_n' the distances of its extremities from B, then the part of P_1 due to v_n is

$$X_n' = -\frac{4v_n\sigma_n}{3b^2} + \frac{4}{3}\int_{\sigma_n'}^{\sigma_n}\frac{v_n x dx}{(\sigma_n-\sigma_n')b^2} = -\frac{2v_n(\sigma_n-\sigma_n')}{3b^2} \tag{C68}$$

Now

$$v_n = \frac{b\mu_n}{c-\rho_n} \qquad \sigma_n = \frac{b^2}{c-\rho_n} \qquad \sigma_n' = \frac{b^2}{c-\rho_n'} \tag{C69}$$

Thus

$$X_n' = -\frac{2\mu_n b(\rho_n-\rho_n')}{3(c-\rho_n)^2(c-\rho_n')} \tag{C70}$$

Adding (C47) and (C48) and summing for all integral values of n from ∞ to 0 gives

$$P_1 = -2\sum_{n=0}^{n=\infty} \frac{\mu_n b(\rho_n - \rho_n')}{(c-\rho_n)^2 (c-\rho_n')} \tag{C71}$$

In order to find the portion P_2 of P due to φ_2 it must be recognized that the original source now is in B. Let σ_n, σ_n' be the distances of the extremities of the nth image in B from B, due to φ_2, then expressing μ_n, σ_n, and σ_n' in terms of v_n, σ_n, and σ_n' gives

$$P_2 = -2\sum_{n=0}^{n=\infty} \frac{v_n(\sigma_n - \sigma_n')}{b^2} \tag{C72}$$

and

$$P = -2\sum_{n=0}^{n=\infty} \frac{\mu_n b(\rho_n - \rho_n')}{(c-\rho_n)^2 (c-\rho_n')} - 2\sum_{n=0}^{n=\infty} \frac{v_n(\sigma_n - \sigma_n')}{b^2} \tag{C73}$$

where μ_n, ρ_n, and ρ_n' refer to the images of A's motion, and γ_n, σ_n, and σ_n' to those of B's motion. We now calculate the following values

$$\begin{aligned} \rho_0 &= 0 & \rho_0' &= 0 \\ \rho_1 &= \frac{a^2 c}{c^2 - b^2} & \rho_1' &= \frac{a^2}{c} \\ \rho_2 &= \frac{a^2 c(c^2 - a^2 - b^2)}{(c^2-b^2)^2 - a^2c^2} & \rho_2' &= \frac{a^2(c^2-a^2)}{c(c^2-a^2-b^2)} \end{aligned} \tag{C74}$$

also, if m_1 is the mass of the liquid displaced by A we obtain

$$\begin{aligned} \mu_0 &= a^2 \dot{a} = \frac{\dot{m}_1}{4\pi} \qquad \mu_1 = \frac{ab\dot{m}_1}{4\pi(c^2-b^2)} \\ \mu_2 &= \frac{a^2 b^2 \dot{m}_1}{4\pi\left\{(c^2-b^2)^2 - a^2c^2\right\}} \end{aligned} \tag{C75}$$

The $v's$ and $\sigma's$ can be obtained by symmetrically interchanging "a" and "b" and putting m_2 for m. If we write M_2, N_2 for the two series in the right-hand side of (C53) we find

$$M_2 = \frac{\dot{m}_1 b}{4\pi c^2}\left[1 + \frac{a^3 b^3}{\left(c^2 - a^2\right)\left(c^2 - a^2 - b^2\right)^2}\right.$$

$$\left. + \frac{a^6 b^6}{\left[\left(c^2 - a^2\right)^2 - b^2 c^2\right]\left\{\left(c^2 - b^2\right)^2 - 2a^2 c^2 + a^2\left(a^2 + b^2\right)\right\}^2} + \cdots\right] \quad \text{(C76)}$$

$$N_2 = \frac{\dot{m}_2 a^3 b}{4\pi c}\left[\frac{1}{\left(c^2 - a^2\right)^2} + \frac{a^3 b^3}{\left(c^2 - a^2 - b^2\right)\left\{\left(c^2 - a^2\right)^2 - b^2 c^2\right\}^2} + \cdots\right] \quad \text{(C77)}$$

and

$$P = -2\left(M_2 + N_2\right) \quad \text{(C78)}$$

We see that M_2 is of the order c^{-2} and N_2 of the order c^{-5}.

The value of the portion of F_2 which depends on the square of the velocity is more difficult to obtain. We will obtain an approximate value for terms with c to exponents to -5. Let $u = a^2\dot{a}$, $\text{v} = b^2\dot{b}$, and let P_n be zonal harmonics when the origin is at A and the axis is BA. Similarly, P_n' are the zonal harmonics when the origin is at B.

Near the surface of B

$$\varphi_1 = \sum_{n=1}^{n=\infty} A_n\left\{R^n + \frac{b^{2n+1}}{(n+1)R^{n+1}}\right\}P_n' + \text{Constant} \quad \text{(C79)}$$

and

$$\varphi_2 = -\frac{\text{v}}{R} + \sum_{n=1}^{n=\infty} B_n\left\{R^n + \frac{b^{2n+1}}{(n+1)R^{n+1}}\right\}P_n' + \text{Constant} \quad \text{(C80)}$$

Omitting the primes for the present and writing C_n for the coefficient of P_n^2 in the expression for φ we obtain

$$Q = 2\int_{-1}^{1}\left(\sum C_n P_n\right)^2 \mu d\mu \quad \text{(C81)}$$

Since P_n^2 is unchanged when $-\mu$ is written for μ we have

$$\int_{-1}^{1} P_n^2 \mu d\mu = 0 \tag{C82}$$

So that

$$Q = 4\sum C_m C_n \int_{-1}^{1} P_m P_n \mu d\mu \tag{C83}$$

where the summation extends to all positive integral values of m, n except for $m = n$. Let

$$\Phi = \int_{\mu}^{1} P_m P_n d\mu \tag{C84}$$

Then

$$\int_{-1}^{1} P_m P_n \mu d\mu = \int_{-1}^{1} P_m P_n d\mu + \int_{-1}^{1} \Phi d\mu = \int_{-1}^{1} \Phi d\mu \tag{C85}$$

From Ferrers [26] we have

$$\Phi = \frac{1}{(m-n)(m+n+1)} \left\{ \frac{n(n+1)}{2n+1} P_m \left(P_{n+1} - P_{n-1}\right) - \frac{m(m+1)}{2m+1} P_n \left(P_{m+1} - P_{m-1}\right) \right\} \tag{C86}$$

Thus $\int_{-1}^{1} \Phi d\mu$ is zero unless $m = n+1$ or $n-1$ and in these cases the integral is (for $m = n+1$)

$$\int_{-1}^{1} \Phi d\mu = \frac{2(m+1)}{(2m+1)(2m+3)} \tag{C87}$$

(for $n-1$)

$$\int_{-1}^{1} \Phi d\mu = \frac{2m}{(2m-1)(2m+1)} \tag{C88}$$

Putting $m = 0$, $r' = R$ into (8) and (10) we have

$$\frac{P_n'}{R^{n+1}} = \frac{P_0}{c^{n+1}} - \frac{(n+1)P_1 r}{c^{n+2}} + \frac{(n+1)(n+2)P_2 r^2}{2!c^{n+2}} - \dots \tag{C89}$$

and

$$\frac{(-)^n P_n}{r^{n+1}} = \frac{P_0'}{c^{n+1}} + \frac{(n+1)P_1'R}{c^{n+2}} + \frac{(n+1)(n+2)P_2R^2}{2!c^{n+2}} - \cdots \quad \text{(C90)}$$

Now if sphere B were absent the value of φ_1 would be

$$\varphi_1 = -\frac{u}{r} \quad \text{(C91)}$$

The value of this near sphere B is

$$\varphi_1 = -\frac{u}{c}\left(P_0' + \frac{P_1'R}{c} + \frac{P_2'R^2}{c^2} + \cdots\right) \quad \text{(C92)}$$

To make the velocity at the surface of B vanish we add the potential φ_1' given by the series

$$\varphi_1' = -\frac{ub^2}{c^2}\left(\frac{P_1'}{2R^2} + \frac{2P_2'b^2}{3cR^3} + \frac{3P_3'b^4}{4c^2R^4} + \cdots\right) \quad \text{(C93)}$$

Transforming each term of the series by using (C69) and (C70) the values of φ_1 near A becomes

$$\varphi_1 = -\frac{u}{r} - \frac{ub^2}{2c^4}\left\{P_0 - \frac{\mathrm{P}_1 r}{c} + \cdots\right\} \quad \text{(C94)}$$

Adding the proper series, the value of φ_1 near A becomes

$$\varphi_1 = -\frac{u}{r} - \frac{ub^3}{2c^4} + \frac{ub^3}{2c^5}\left(r + \frac{a^3}{2r^2}\right)P_1 + \cdots \quad \text{(C95)}$$

The added term produces at B a constant term of the order c^{-7}, which is negligible, hence the value of $\varphi_.$, near B is

$$\varphi_1 = -\frac{u}{c} - \frac{u}{c^2}\left(R + \frac{b^3}{2R^2}\right)P_1' - \frac{u}{c^3}\left(R^2 + \frac{2b^5}{3R^3}\right)P_2' + \cdots \quad \text{(C96)}$$

Changing P_1 to $-P_1'$ it follows from (C75) that the value of φ_2 near B is

$$\varphi_2 = -\frac{\mathrm{v}}{R} - \frac{u}{c} - \frac{\mathrm{v}a^3}{2c^4} - \left(\frac{u}{c^2} - \frac{\mathrm{v}a^3}{2c^5}\right)\left(R + \frac{b^3}{2R^2}\right)P_1' - \frac{u}{c^3}\left(R^2 + \frac{2b^5}{3R^3}\right)P_2 - \cdots \tag{C97}$$

Substituting $R = b$ gives

$$\varphi_2 = -\frac{\mathrm{v}}{b} - \frac{u}{c} - \frac{\mathrm{v}a^3}{2c^4} - \frac{3b}{2}\left(\frac{u}{c^2} + \frac{\mathrm{v}a^3}{2c^5}\right)P_1' - \frac{5\varphi b^2}{3c^3}P_2' - \cdots \tag{C98}$$

Also

$$\left(\frac{d^3\varphi}{dR^3}\right)_b = -\frac{2\mathrm{v}}{b^3} - \frac{3}{b}\left(\frac{u}{c^2} + \frac{\mathrm{v}a^3}{2c^5}\right)P_1' - \frac{10u}{c^3}P_2' - \cdots \tag{C99}$$

Thus

$$Q = 2\int_{-1}^{1}\varphi^2\mu d\mu = 4\left(\frac{u\mathrm{v}}{c^2} + \frac{u^2 b}{c^3} + \frac{\mathrm{v}^2 a^3}{c^5}\right) + \frac{8u^2 b^2}{3c^5} \tag{C100}$$

Also, using (C42) we have

$$\frac{d^2Q}{dR^2} = 4\int_{-1}^{1}\varphi\frac{d^2\varphi}{dR^2}\mu d\mu = 8\left\{\frac{2u\mathrm{v}}{b^2c^2} + \frac{u^2}{bc^3} + \frac{\mathrm{v}^2a^3}{b^2c^5} + \frac{8u^2b}{3c^5}\right\} \tag{C101}$$

Putting back the values of u and v gives

$$\frac{b^2}{4}\frac{d^2Q}{dR^2} - \frac{1}{2}Q = a^2b^2\left(\frac{2\dot{a}\dot{b}}{c^2} + \frac{ab^2\dot{b}^2}{c^5}\right) \tag{C102}$$

By (C76), (C77), and (C78)

$$P = -\frac{2a^2b\dot{a}}{c^2} - \frac{2a^3b^3\dot{b}}{c^5} - \textit{higher powers of } 1/c \tag{C103}$$

Using (C66) the part of the force $F_{2\mathrm{v}}$ depending on the square of the velocity is

$$F_{2\mathrm{v}} = -\frac{\pi\rho a^3b^4\dot{b}^2}{c^5} \tag{C104}$$

and is seen to vary inversely with c^5. Finally we have

$$F_2 = -2\pi\rho b^2\frac{d}{dt}\left(M_2 + N_2\right) - \frac{\pi\rho a^3b^4\dot{b}^2}{c^5} \tag{C105}$$

The value of F_1, the force on A towards B, is obtained by symmetrically interchanging "a" and "b".

Neglecting all powers of $1/c$ above the second we have

$$F_2 = -\frac{2\pi\rho b^2}{c^2}\frac{d}{dt}\left(a^2 b\dot{a}\right) \tag{C106}$$

Letting

$$\begin{aligned} a &= \bar{a} + \alpha \sin\frac{2\pi t}{T} \\ b &= \bar{b} + \beta \sin\frac{2\pi}{T}(t-\varepsilon) \end{aligned} \tag{C107}$$

where $\bar{a}$ and $\bar{b}$ are the mean radii of the spheres α and β are the (half) amplitudes of the oscillations, T is the period the oscillation and ε is the angle the oscillation of B lags A, then the average values of F_2, i.e., $\overline{F_2}$ is

$$\begin{aligned} \overline{F_2} &= -\frac{2\pi\rho}{Tc^2}\int_0^T b^2 \frac{d}{dt}\left(a^2\dot{a}b\right)dt = \frac{4\pi\rho}{Tc^2}\int_0^T a^2 b^2 \dot{a}bdt \\ &= \frac{16\pi^3\rho}{Tc^2}\left(\overline{ab}\right)^2 \alpha\beta \int_0^T \cos\frac{2\pi t}{T}\cos\frac{2\pi}{T}(t-\varepsilon)\,dt \\ &= \frac{8\pi^3\rho\bar{a}^2\bar{b}^2\alpha\beta}{Tc^2}\cos\frac{2\pi\varepsilon}{T} = \overline{F} \end{aligned} \tag{C108}$$

If the spheres are breathing out then in at the same time, that is if they are in phase $(\varepsilon = 0)$ then the force is a maximum value and it is a force of attraction. When 180° out-of-phase the force is a maximum and it is a force of repulsion. If the phase differs less than a quarter of a period they attract each other. If the phase differs more than a quarter and less than three quarters of a period they will repel each other.

The maximum force between two identical spheres of radius "a" and vibratory half amplitude "α" is

$$F_{\max} = \frac{8\pi\rho a^4\alpha^2}{Tc^2} \tag{C109}$$

Appendix D: Entropy, Heat Engines, and Neutrinos

A heat engine consists of thermodynamic system (usually a gas), a mechanical means of expanding and compressing the gas, and a means of adding and removing heat. Figure Dl shows a very simple thermodynamic system (a single particle), Figure D2 shows the system with a mechanical enclosure (the cylinder and piston), Figure D3 shoes a simplified heat transfer device, and Figure D4 shows the heat engine.

The ball (i.e., the thermodynamic system) is frictionless and perfectly elastic, the piston and cylinder are perfectly elastic and frictionless, and the heat transfer device always impacts the system “head-on” so that no transverse force is applied to the rod on which it translates back-and-forth.

The thermodynamic system can oscillate back and forth between the cylinder cap and the piston head, the ball can be slowed-down or speeded-up by controlling the rotation rate of the heat transfer device, and the piston can move to the right or left depending upon the value of F.

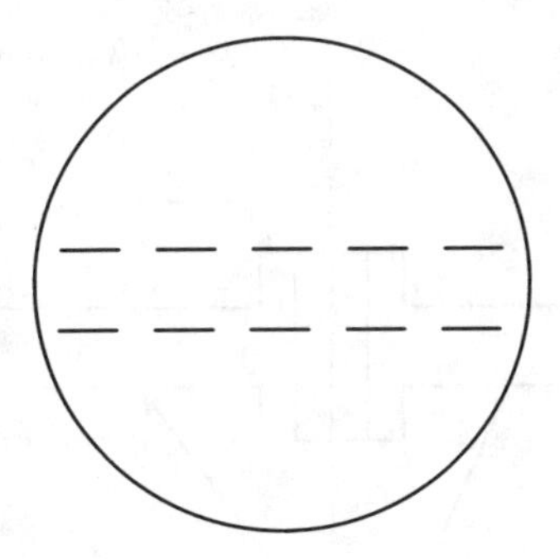

Figure D1. Thermodynamic System
(One Ball)

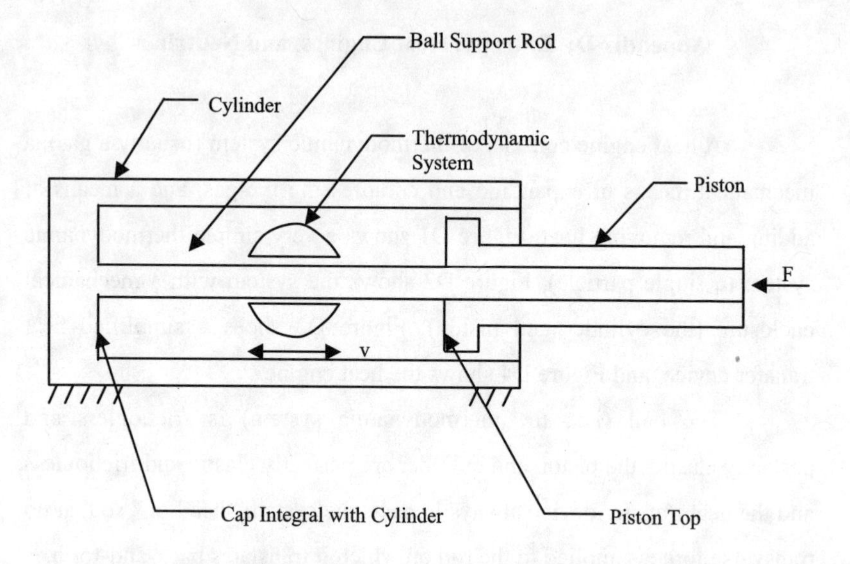

Figure D2: Thermodynamic System and Its Enclosure (Thermodynamic Device)

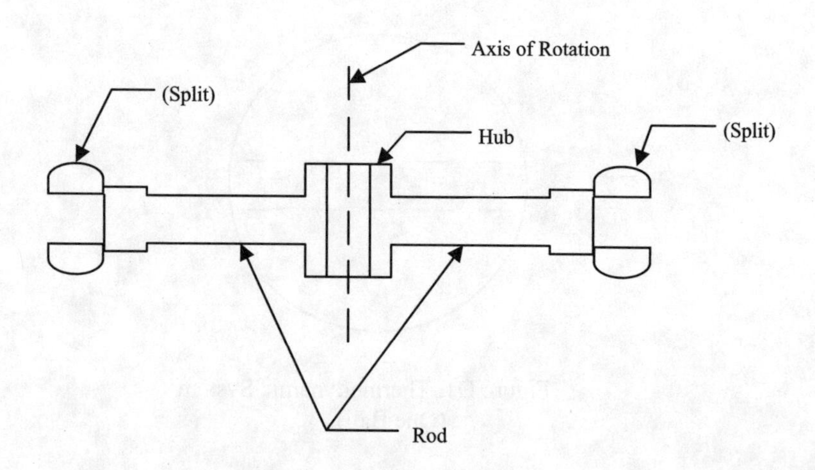

Figure D3. Heat Transfer Device

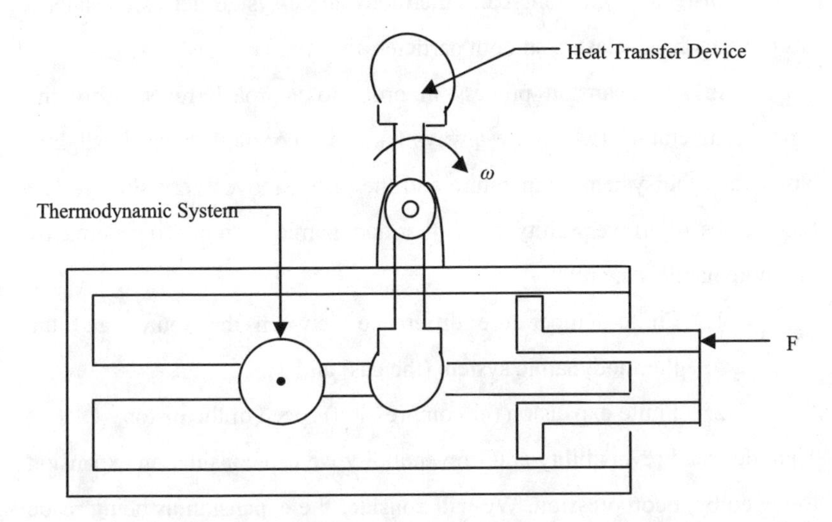

Figure D4. Heat Transfer Device Mounted to a Piston and Cylinder

The heat engine is a device for interchanging mechanical and thermal energy. A heat engine experiences a "process" when the "system" changes from one value of thermal energy (i.e., kinetic energy of oscillatory motion) and volume to another thermal energy – volume value. Each thermal energy – volume value is a "state". A process can approach reversibility if it meets the two requirements:

1. The system temperature and the source temperature have approximately the same values (i.e., their impact velocities are approximately the same magnitude, sense, and direction.)
2. The piston velocity during expansion or compression approaches zero.

The two processes of primary theoretical interest are:

1. Expansion, or compression, with no heat transfer (adiabatic process).
2. Expansion, or compression, with constant temperature (i.e., isothermal), requires heat transfer to or from the source.

Instead of the "one-ball" thermodynamic system let us consider a gas of many hard, elastic, smooth particles, the "perfect" gas.

In an expansion process in order to approach reversibility the source temperature (which we always take as a constant value) is slightly larger than the system temperature and the piston moves very slowly. The two causes of irreversibility for a thermodynamic system, considered by thermodynamicists, are:

1. Finite temperature difference between the source and the thermodynamic system (the gas) and
2. Finite expansion (or compression) speed of the piston.

To understand reversibility and irreversibility we can consider an expansion followed by a compression. We will consider the expansion as being made up of two processes. The main reason for using two processes in the analysis of heat engines is that we have a heat source which remains at constant temperature and it supplies heat to the engine as the piston expands and the most work can be done if the heat source and the system are approximately at the same temperature. This process is followed by an adiabatic process in order to decrease the system temperature. On the return compression a constant temperature process permits heat to be dumped from the gas (which is colder than it was during the isothermal expansion) to the heat sink. This is followed by an adiabatic compression to return to the initial condition, if all the processes were reversible. Returning to the original condition completes one "cycle" of the heat engine.

First we expand using a constant temperature process and we follow this with a no-heat transfer to or from the system, which is an adiabatic expansion. (Again, we emphasize that the system is the gas.) In designing a heat engine the piston stroke is quite constrained and often specified. Now, by having the two processes any end-point conditions can be obtained between the complete isothermal expansion and the complete adiabatic expansion by combining the processes differently. Figure D5 shows the pressure, p, versus the specific volume, v, for the two different

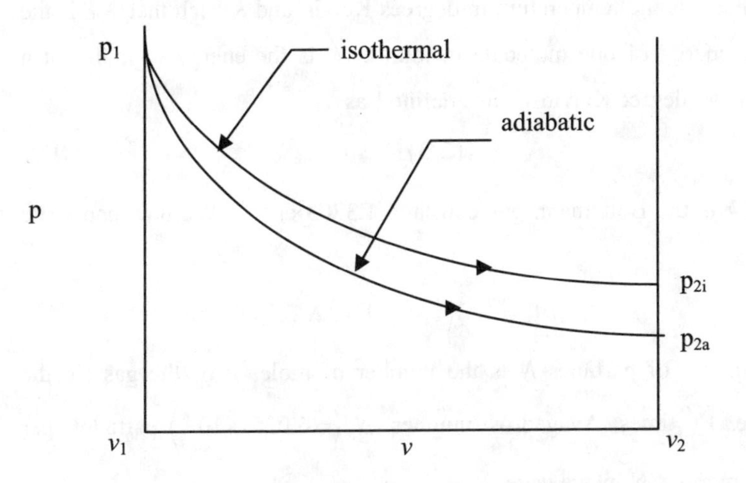

Figure D5. Isothermal and Adiabatic Expansions

processes. Now by using a long isothermal expansion followed by a short adiabatic expansion an end-point pressure just below p_{2i} will be reached. Alternatively, if a short isothermal stroke is followed by a long adiabatic stroke an end-point just above p_{2a} will be reached.

Let us now obtain the equations for describing the isothermal expansion and the adiabatic (no-heat-transfer) expansion process. First, we need an "equation of state". The equation of state for the ideal gas is simply the pressure, p, density, ρ, thermal velocity, v_r, relation

$$p = \frac{1}{3}\rho \mathrm{v}_r^{\,2} \qquad \text{(D1)}$$

This can be written in terms of temperature as follows

$$p = \frac{1}{3}\rho \mathrm{v}_r^{\,2} = \frac{1}{3}\frac{Nm}{V}\mathrm{v}_r^{\,2} = \frac{2}{3}\frac{N}{V}\left(\frac{1}{2}m\mathrm{v}_r^{\,2}\right) \qquad \text{(D2)}$$

where N is the total number of particles (all assumed to be alike with mass, m, and contained in volume V). Define the temperature T and a constant K

such that T is the temperature in degrees Kelvin and K such that KT is the kinetic energy of one molecule in joules. (K is the energy in joules of a particle per degree Kelvin.) Now, define k as

$$k = (2/3)K \tag{D3}$$

where k is the Boltzmann gas constant, 1.3803×10^{-23}. We now can write (D2) as

$$pV = \tfrac{2}{3}N\left(\tfrac{1}{2}m\mathrm{v}_r^{\,2}\right) = \tfrac{2}{3}NKT$$

The number of particles N is the number of moles n of the gas (in the volume V) times Avagadros number $N_o\left(=6.0251\times10^{26}\right)$ particles per kilogram mole. Now we have

$$pV = (2/3)nN_0KT = n\left[N_0(2/3)K\right]T = \left(nN_0k\right)T \tag{D4}$$

Most physicists and chemists define the gas constant R_{pc} as

$$R_{pc} = kN_o \tag{D5}$$

so that the gas equation of state is

$$p\mathrm{v} = nR_{pc}T \tag{D6}$$

Engineers generally define the gas constant R_e as

$$MR_e = nR_{pc} \tag{D7}$$

so that the gas equation is

$$PV = MR_eT \tag{D8}$$

We will use (D8) except that we generally divide both sides of the equation by M and use the specific volume v which is

$$v = \frac{V}{M} \tag{D9}$$

then

$$p\mathrm{v} = R_eT \tag{D10}$$

For the isothermal reversible process going from state 1 to state 2 with $T_{so} = T_g$ where T_{so} is the source absolute temperature and T_g is the gas temperature, the energy balance is

$$q = \mathrm{w}_{1\to 2} \tag{D11}$$

since the internal energy is constant. In this expression $q_{1,2}$ is the heat transferred per unit mass of the gas and $\mathrm{w}_{1\to 2}$ is the work done by the gas going from state 1 to state 2. The work is given by

$$\mathrm{w}_{1\to 2} = \int_1^2 pdv = \int_1^2 \frac{R_e T}{v} dv = R_e T \ln \frac{v_2}{v_1} \tag{D12}$$

Consider now the adiabatic reversible process from state 2 to state 3. The energy balance is

$$q_{2\to 3} = \int \delta q = \int (du + d\mathrm{w}) \tag{D13}$$

Now $\delta q = 0$ and

$$d\mathrm{w} = pdv \tag{D14}$$

so

$$du = -\, pdv \tag{D15}$$

Further, the specific heat at constant volume "c_v" is given by

$$c_v = \frac{du}{dT} \tag{D16}$$

Where u is the internal energy of the gas. Thus

$$du = c_v dT = -\, pdv \tag{D17}$$

The energy balance can be written in terms of the enthalpy h where

$$h = u + pv \tag{D18}$$

and

$$dh = du + pdv + vdp \tag{D19}$$

or

$$du + pdv = dh - vdp \tag{D20}$$

Thus

$$\delta q = du + pdv = dh - vdp \tag{D21}$$

The specific heat at constant pressure "c_p" is given by

$$c_p = \frac{dh}{dT} \tag{D22}$$

or

$$dh = c_p dT \tag{D23}$$

Now, using (D21) and (D23) with $\delta q = 0$ we obtain

$$c_p dT = vdp \tag{D24}$$

Dividing (D24) by (D16)

$$\frac{c_p dT}{c_v dT} = -\frac{vdp}{pdv} \tag{D25}$$

Using

$$k = \frac{c_p}{c_v} \tag{D26}$$

gives

$$k = -\frac{v}{p}\frac{dp}{dv} \tag{D27}$$

Integrating from state 2 to state 3 gives

$$\int_2^3 k\frac{dv}{v} = -\int_2^3 \frac{dp}{p} = k \ln\frac{v_3}{v_2} = \ln\left(\frac{v_3}{v_2}\right)^k = \ln\frac{p_2}{p_3} \tag{D28}$$

Now, we have

$$p_2 v_2^{\,k} = p_3 v_3^{\,k} = pv^k = c \tag{D29}$$

We can now evaluate the work integral

$$w_{2\to3} = \int_2^3 p\,dv = \int_2^3 \frac{c\,dv}{v^k} = \left[\frac{cv^{-k+1}}{-k+1}\right]_2^3 = \frac{c}{k-1}\left[\frac{1}{v_2^{\,k-1}} - \frac{1}{v_3^{\,k-1}}\right]$$

$$= \frac{p_3 v_3}{k-1}\left[\frac{1}{v_2^{\,k-1}} - \frac{1}{v_3^{\,k-1}}\right] \tag{D30}$$

Let us now show how to begin at state p_1, v_1 and go to any state specified by the specific volume and pressure between p_{2a} and p_{2i} in Figure D5 by using an isothermal expansion followed by an adiabatic expansion. Let the states be designated as p_1, v_1 at the beginning, p_2, v_2 at the juncture of the two processes, and p_3, v_3 be the ending state, see Figure D6. For the isothermal expansion

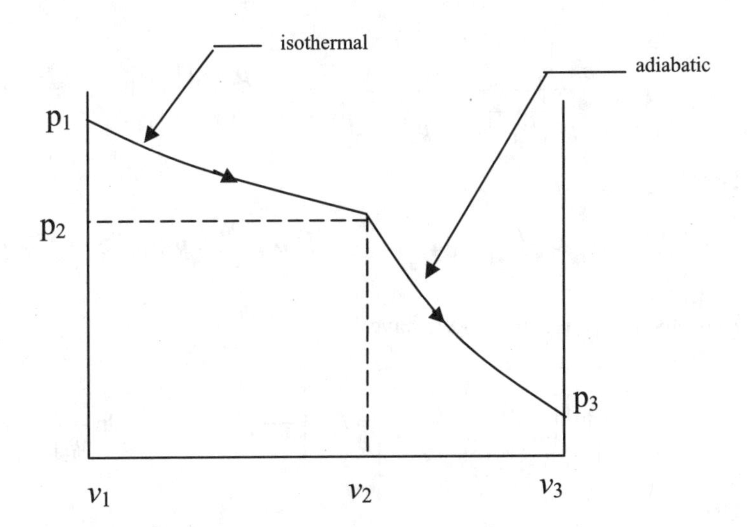

Figure D6. An Isothermal Expansion Followed by Adiabatic Expansion

we have

$$pv = R_e\,T = p_1 v_1 \tag{D31}$$

and for the adiabatic expansion we have

$$pv^k = p_3 v_3{}^k \tag{D32}$$

At the juncture we equate the pressures and the specific volumes. From (D31)

$$p_2 = p_1 v_1 / v_2 \tag{D33}$$

Substituting into (D32) we have

$$\frac{p_1 v_1}{v_2} v_2{}^k = p_3 v_3{}^k \tag{D34}$$

or

$$v_2 = \left(\frac{p_3}{p_1}\right)^{\frac{1}{k-1}} \frac{v_3{}^{\frac{k}{k-1}}}{v_1{}^{\frac{1}{k-1}}} \tag{D35}$$

and

$$p_2 = \frac{p_1 v_1 p_1{}^{\frac{1}{k-1}} v_1{}^{\frac{1}{k-1}}}{p_3{}^{\frac{1}{k-1}} v_1{}^{\frac{k}{k-1}}} = \frac{p_1{}^{\frac{k}{k-1}}}{p_3{}^{\frac{1}{k-1}}} \frac{v_1{}^{\frac{k}{k-1}}}{v_3{}^{\frac{k}{k-1}}} = p_1 \left(\frac{p_1}{p_3}\right)^{\frac{1}{k-1}} \left(\frac{v_1}{v_3}\right)^{\frac{k}{k-1}} \tag{D36}$$

Now

$$q_{1,2} = \Delta\mu_{1,2} + \mathrm{w}_{1\to 2} = 0 + \int_1^2 p dv = \int_1^2 R_e T \frac{dv}{v} = R_e T \ln \frac{v_2}{v_1} \tag{D37}$$

In terms of p_1, v_1, p_3, v_3 we have

$$q_{1,2} = R_e T \ln \left[\left(\frac{p_3}{p_1}\right)^{\frac{1}{k-1}} \left(\frac{v_3}{v_1}\right)^{\frac{k}{k-1}} \right] = R_e T \left[\frac{1}{k-1} \ln \frac{p_3}{p_1} + \frac{k}{k-1} \ln \frac{v_3}{v_1} \right] \tag{D38}$$

Using

$$c_p - c_v = R_e \tag{D39}$$

we have

$$q_{1,2} = T\left[c_v \ln\frac{p_3}{p_1} + c_p \ln\frac{v_3}{v_1} \right] \tag{D40}$$

Equation (D40) shows that the heat per unit of temperature in a reversible path at constant temperature is equal to a function of the beginning and end states only and is not dependent upon the particular path. The paths are limited to an isothermal path followed by an adiabatic path and such that the final end state is between the extremes of the pure isothermal path and the pure adiabatic path, see Figure D5. The heat per unit temperature is $q_{1,2}/T$.

Any monotonically decreasing path can be broken down into a series of isothermal-adiabatic steps, even differential ones and in that case the ratio would be dq/T. It can be proven that for any reversible path the expression dq/T is given by $\left[c_v \ln\frac{p_3}{p_1} + c_p \ln\frac{v_3}{v_1} \right]$. Thus dq/T is a "property" of ideal gases which is called the (differential) entropy "ds". Thus, based on (D40) we can write

$$\begin{aligned} s_1 &= c_v \ln p_1 + c_p \ln v_1 \\ s_3 &= c_v \ln p_3 + c_p \ln v_3 \\ s &= c_v \ln p + c_p \ln v \end{aligned} \tag{D41}$$

Also, we have the equation

$$ds = \frac{dq_{rev}}{T} \tag{D42}$$

in which the symbol dq_{rev} means that the heat is added reversibly. Since "s" is a property its value is determined solely by the thermodynamic variables and not by any path parameters. Thus "s" is not determined by reversibility effects.

Let us now return to the expansion analysis followed by a compression along the same path. The work done in the isothermal portion is

$$\mathrm{w}_{1\to 2} = R_e T_g \ln \frac{v_2}{v_1} \qquad \text{(D43)}$$

where T_g is the gas temperature which is taken to be the same as the source temperature T_{so}. For the adiabatic portion the work is

$$\mathrm{w}_{2\to 3} = \frac{p_3 v_3}{k-1}\left[\frac{1}{v_2^{\,k-1}} - \frac{1}{v_3^{\,k-1}}\right] \qquad \text{(D44)}$$

The heat added is

$$q_{1\to 2} = R_e T_g \ln \frac{v_2}{v_1} \qquad \text{(D45)}$$

which is the same as the work term $\mathrm{w}_{1\to 2}$. The internal energy change all occurs in the process from state 2 to 3 and it is

$$u_{2,3} = -\frac{p_3 v_3}{k-1}\left[\frac{1}{v_2^{\,k-1}} - \frac{1}{v_3^{\,k-1}}\right] \qquad \text{(D46)}$$

and its decrease is the same as the work done from state 2 to state 3. In the reversible compression along the same path each of the above four energies (given by (D43), (D44), (D45), and (D46)) are reversed.

Let us now consider irreversible paths. If $T_{so} > T_g$ then the adiabatic compression must continue until the gas temperature exceeds the source temperature in order that the isothermal transfer of heat from the gas to the source will take place, see Figure D7. Thus net work ($\int pdv$) is required to return to the original state than for the case where the expansion had been done with $T_g = T_{so}$ (since this net work is zero).

Let us now consider a finite piston speed u, and $u \ll \mathrm{v}_r$, where v_r is the rms velocity of the gas. Figure D8 shows the piston, cylinder, and gas. The internal energy of the gas per unit mass is $(1/2)\rho v_{rso}^{\;2}$, where v_{rso} is the rms velocity of the gas when initially at the source temperature. When the piston moves at velocity u the average flow velocity of the gas is $u/2$ and its flow energy is $(1/8)\rho u^2$.

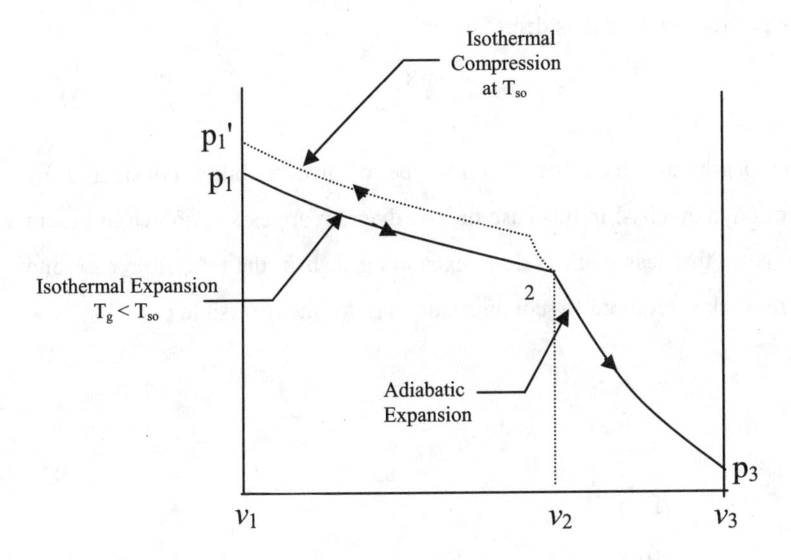

Figure D7. Expansion followed by Compression when $T_{so} > T_g$

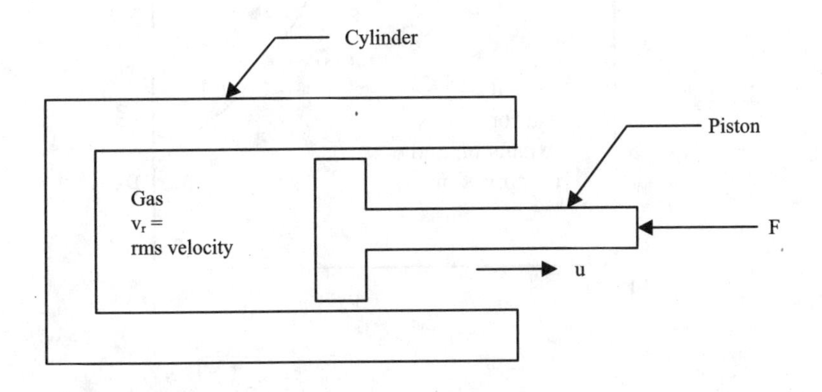

Figure D8. Piston Moving at Velocity u.

This results in a thermal energy decrease. The resulting thermal energy is

$$(1/2)\rho \mathrm{v}_r^{\,2} = (1/2)\rho \mathrm{v}_{rso}^{\,2} - (1/8)\rho u^2 \tag{D47}$$

or

$$\mathrm{v}_r^{\,2} = \mathrm{v}_{rso}^{\,2} - u^2/8 \tag{D48}$$

The gas temperature thus decreased to

$$T = \frac{\mathrm{v}_{rso}^{\,2} - u^2/8}{\mathrm{v}_{rso}^{\,2}} T_{so} \tag{D49}$$

This decrease causes the second type of irreversibility considered by thermodynamicists. In this case the *p-v* diagram appears as shown in Figure D9. Note that less work is done expanding than in the reversible case and more work is required for compression than for the reversible case.

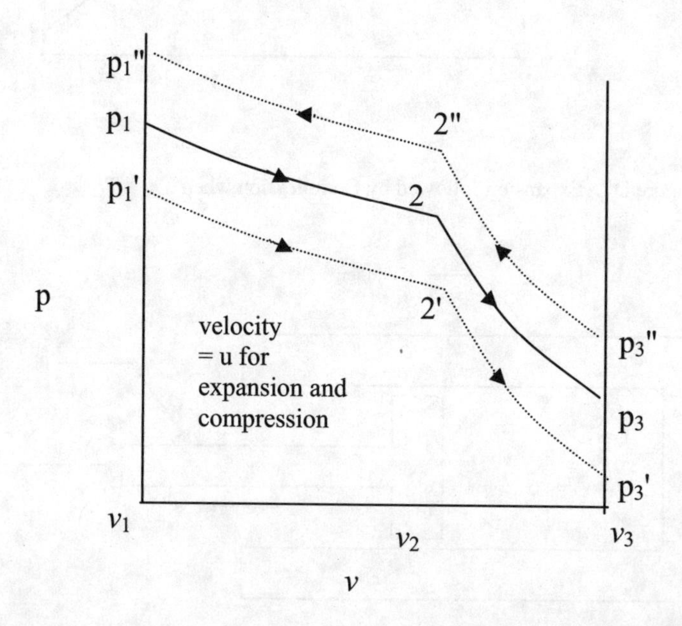

Figure D9. *p-v* Diagram for Expansion and Compression at Finite Piston Velocity u

The entropy change in the reversible expansion compression cycle is

$$\Delta s_{rw} = (\Delta s)_{\text{expansion}} + (\Delta s)_{\text{compr}} \tag{D50}$$
$$= (s_3 - s_1) + (s_1 - s_3) = 0$$

The entropy change in the irreversible expansion-compression cycle is

$$\begin{aligned} \Delta s_{irrev} &= (s_2 - s_1) + (s_1' - s_2') \text{ for temp. mis-match (Figure D7)} \\ &+ (s_3' - s_1') + (s_1'' - s_3'') \text{ for finite piston velocity (Figure D9)} \end{aligned} \tag{D51}$$

Using (D41) the entropy change can be computed. The result is a positive value. The change in entropy times the source temperature is the heat lost beyond the heat required for the reversible cycle.

The reversible heat engine cycle consisting of an isothermal expansion, followed by an adiabatic expansion, followed by an isothermal compression, then an adiabatic compression returning to the initial state is named the Carnot cycle after is discoverer. The *p-v* diagram for the Carnot cycle is shown in Figure D10. The cycle most often is shown

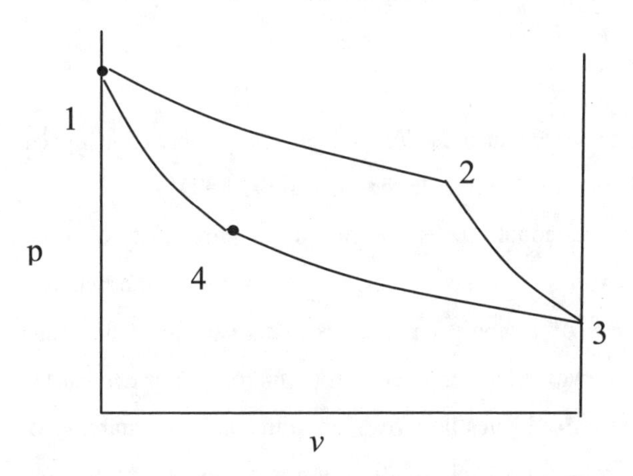

Figure D10. *p-v* Diagram for the Carnot Cycle

as a *T-S* plot since it is rectangular, see the solid lines in Figure D11. Processes 1 to 2 and 3 to 4 are constant temperature and processes 2 to 3 and 4 to 1 are constant entropy (since $dq = Tds$, and dq is zero). The temperature T_{si} is

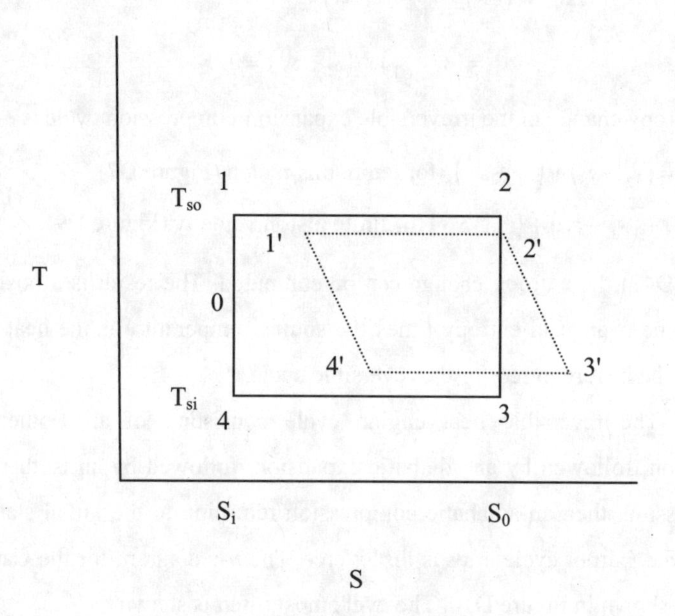

Figure D11. *T-S* Diagram for the Carnot Cycle

the "cold" sink to which heat is dumped on the compression cycle. An irreversible cycle is shown by the dotted lines in Figure D11.

In all macroscopic cycles the entropy is universally observed to increase. Entropy forms the basis for design decisions in all thermal systems and in the analysis of biological processes. Microscopic cycles, such as molecular orbits, however have neither entropy increases nor decreases.

The theory of physics here rests upon the assumed stability of the neutrino. The entropy for a cycle of its action is computed by determining the entropy of the ambient "gas" which is ready to flow into the neutrino then compute the entropy of that same "gas", which by then has changed to a solid, as it exits from the neutrino. The entering entropy is

$$s_1 = c_v \ln p_1 + c_p \ln v_1 \quad \text{(D52)}$$

The exiting entropy is

$$s_2 = c_v \ln p_2 + c_p \ln v_2 \quad \text{(D53)}$$

At the exit the pressure is zero, since the particles are all aligned. Further, the specific volume (volume per unit mass) is essentially zero. Thus, $s_2 \approx 0$ and $s_2 - s_1$ is negative. Thus the neutrino, as it is envisioned here, is an entropy reducer.

Contemporary physics rejects the possibility that such a state as envisioned for the neutrino is possible. However, the arguments presented in this book make the neutrino state seem plausible, but certainly do not prove it is possible. On the other hand contemporary physics assumes that all the organization was placed in a minute ball which has been expanding for ten billion years and provides the organization observed as it is expanding. When the kinetic particle theory of physics is judged against the expanding ball the kinetic particle theory sounds better!!

Appendix E: A Personal History of the Kinetic Particle Unified Physics

Many of my fellow members of the American Physical Society will look at this book, possibly glance at the simple set of postulates, then immediately dismiss the book. They would immediately think it impossible to derive all the complexities of physics starting with just one "billiard ball" type of particle. Recently I prepared a paper, which essentially was the contents of this book, and submitted it to several physics journals for their review. I feel safe in saying that none of the reviewers read the paper carefully enough to understand its content.

Such a reception of a new structure of physical science, even when it better explains the way the universe works, is normal. Thomas Kuhn in his book The Structure of Scientific Revolution [21] calls a new comprehensive theory a paradigm. In his book he explores several different paradigms throughout the history of science. The general pattern is that the old scientists cling to their paradigm until they die and the following generation adopts the new theory.

This may be my last writing in physics, I was 76 years old on January 9, 2004, so I thought it worthwhile to document what I remember[1] about the development of this theory. I will first start by giving some of my background before starting to work on the theory.

I was born in the rural community of Gatewood near Fayetteville, West Virginia. My parents were Paul Alexander Brown and Mary Lou Fox Brown. I have an older brother Paul Jerry Brown born July 28, 1925

[1] I could write a book entitled *What Little I Remember* like Frisch [24] wrote. However, my book probably would be much shorter than his.

and a younger sister Mary Jane Brown Songer, born November 13, 1929. We lived on a 10 acre plot and had a garden, chickens, and a milk cow. My father worked in the coal mines during the evenings and repaired automobiles and built houses during the days. We went to the Church of the Brethren nearly every Sunday. We were very poor during the depression years, 1930 to 1940.

We ran throughout the neighborhood and played most of these depression years. We spent a lot of time running at full speed playing fox and hounds. The West Virginia hills were beautiful in the summer, and in the winter, as well as in the spring and fall. One of our best toys often was a pot lid which we used as our automobile (actually just the steering wheel). There was a story of a boy who lived in the town of Fayetteville 3 miles away who "drove" his lid car to our school every morning for one whole school year. Turned around by backing his car down the lane beside our school, and of course looking over his right shoulder as he was shifting into reverse steering his car and backing down, stopping, then shifting into first gear and driving back to Fayetteville.

I attended Victory Grade School, a 3 room 3 classes per room facility, until graduation in 1940. I read all the books I could find, which were about 4 or 5 school books a year. I then attended Fayetteville High School, graduating at 15 years of age in 1943. Throughout the first 13 grades of school I excelled in grammar, history, and mathematics – anything where logic and memory were useful.

I studied engineering the next 9 years first at West Virginia Tech, then West Virginia University (obtaining a BS in 1946 and a BSME in 1947) then at Purdue University (obtaining an MSME degree in 1950,[1]and a PhD in 1952).

---

[1] There was a semester interruption of my Purdue stay by a fall 1947 semester at Cornell University.

I was at West Virginia Tech during wartime and Professor "Daddy Boy" Smith was the only math teacher. There were 20 math students taking approximately 10 different courses each semester. There was no way he could lecture on so many subjects so the first day of class he made certain we had the correct book. He then told us to work the odd problems and submit them for credit in the course. He told us to come to his office if we needed help. I soon found he could not help me and I had to "learn-on-my-own". I self-taught five courses including calculus. I was fascinated by the infinite processes of calculus. To better learn the subject, I re-took calculus when I transferred to West Virginia University. One highlight at West Virginia University was the final lecture by Professor Ford in physics. He explained the quantum levels of atoms – how one electron jumps from one orbit to another emitting, or absorbing a photon. It was fascinating and believable.

My graduate courses were equally divided among mathematics, mechanics, and mechanical design. My background in mechanical design, engineering mechanics, and the mathematics required for modeling mechanical systems was the ideal background for developing a mechanical model of the universe.

During my school years there were several phenomena which struck me as extremely profound:

1. How does a lightning bug produce light?
    - A pre-school/kindergarten question
2. Why do iron filings on a paper above a magnet make their characteristic patterns?
    - A sixth grade demonstration
3. The mathematical modeling the lever and trusses was a fascinating joining of mathematics and physics.
4. I heard something from a fellow student at Cornell University about Einstein's mass growth or matter shortening. I was astounded and unbelieving, but I don't remember any details.

5. During my school career (as well as before and even after) I have often dreamed of personally overcoming gravity. I often stood unsupported above the ground while people were around – but they never noticed even when I told them. Also, I could run by taking long and longer strides, often 100 meter strides.

Soon after leaving Purdue I moved to California and met an American Airlines stewardess Miss Jimmie Marie Dikeman from Oolagah, Oklahoma. We were soon married, January 18, 1952, and begat four children, Paul Alexander Brown II was born January 18, 1953; Stephanie Ann Brown was born June 27, 1956; Barbara Louise Brown was born October 1, 1959; and Carolyn Jane Brown was born July 11, 1962.

I spent 19 years in aircraft structural analysis, aerospace design management, and physics research at various aerospace companies from 1951 through 1970. I started teaching engineering mechanics part-time at West Virginia University in 1947 and continued at the University of Southern California, and UCLA. I began teaching design, thermodynamics, and engineering mechanics full-time at Mississippi State University and continued for 21 years from 1970 thru 1991. The last 13 years I have spent approximately half time doing research.

In 1958 while I was a first level manager at Hughes Aircraft Company I got the idea that gravitation was produced by ether particles in space somehow pushing bodies together. In 1963 while I was a second level manager at the Aerospace Corporation in El Segundo, California I stumbled across the idea that gravitation was caused by a mutual shielding of one body be another and that the ether particles pushed the bodies together. (Later I discovered the idea was fallacious. Gravitation is caused by matter producing a distant expansion-contraction wave in the ether and that lets the ether push the bodies together.) I worked on the ether particle theory until 1967 at the Aerospace Corporation and finally evolved the concept that everything was made up the same particles that make up the ether. I published 3 books in this period of time: (References [13], [14], and [15])

1. *Unified Physics Part 1* JMB Co, Box 45004, Los Angeles, CA 90045, 1965.
2 *Advanced Physics*, JMB Co, Box 153, Los Angeles, CA 90045, 1966.
3. *Advanced Physics, 3rd Edition*, JMB Co, Box 153, Los Angeles, CA 90045, 1967.

These books had many fundamental errors but I felt they would advance the development of the kinetic particle theory of physics.

While at the Aerospace Corporation in 1966 I requested support for my efforts from the president of the corporation, the late Dr. Ivan Getting. He formed a panel, chaired by Dr. Jack Irving, to review my work. (The Corporation had recently sent Dr. Irving to the California Institute of Technology to obtain his PhD.) I made a presentation which I felt all members of the panel would feel the world surely must be made up of kinetic particles. They didn't. As I left the presentation one of the panel members Dr. Robert Huddelstone remarked that it was unbelievable that I would sacrifice my career as an engineering manager (I was an associate department head at the time) on a fantasy. I thought he would soon see I was on the right track. In the presentation Dr. Irving asked me if I wanted to study physics at Cal Tech (as I remember). I said, I wanted to continue working physics part time. In later years I either heard, or deduced, that Dr. Getting was willing to fund my education to obtain a PhD in physics. If so, I still would not have accepted the offer primarily because I thought the physics I would be taught would be incorrect-particularly the relativity theory.

The third book I had published *Advanced Physics – 3rd Edition* addressed more than physics. It was my first step toward organizing all of science. Even with its severe limitations it came into the hands of Dr. Robert M. Wood the deputy director of Research at Douglas Aircraft Company, in Santa Monica, California. It lead me into a job wherein I could develop my theory of physics, which I so ardently wanted.

Dr. Wood believed that unidentified flying objects (UFOs) were spacecraft built and manned by extra-terrestrials. Because of the great distances to other stars he believed that the vehicles could travel at speeds orders of magnitude greater than the speed of light. Furthermore, he had great confidence in his ability. "If extraterrestrial's can build a flying saucer, so can I."

Dr. Wood had an associate, Dr. Darell B. Harmon, Jr., Dr. Harmon was the inventor and subsequent manager of a major project at the Douglas Aircraft Company to develop "external burning" for greatly enhancing maneuverability of anti-ballistic missile defense rockets. Dr. Harmon held beliefs about UFO's similar to Dr. Wood.

Dr. Wood interviewed me for employment and asked what I believed regarding a list of some dozen subjects. One of the subjects was UFOs. My answer was that I believed they were real but natural atmospheric disturbances. Dr. Wood believed that UFO's were powered by anti-gravity propulsion and I was researching gravity so he hired me in spite of my disbelief in extraterrestrials. Thus the Wood-Harmon-Brown team was formed to invent a new propulsion system based on obtaining an understanding of the mechanism of gravity.

Well here we were, our leader Bob Wood, and our consultant Darell Harmon, both of whom worked full-time at other tasks, and I, who was promised at least half of my time to work on my theory and the remainder to work on Bob's projects, whatever they may be. All three of us had been very satisfied with our engineering endeavors. All of us had the best backgrounds possible to develop a mechanical theory of physics. All three of us rejected the tenets of relativity theory (I thought all the known relativity experiments could be derived from classical theory and Bob and Darell thought super-optic travel was possible), and we had the vast resources of Douglas Aircraft Company at our disposal. We were ready to go. We hired an experienced and well trained mathematical physicist, Dr.

Leon A. Steinert just to make sure we had all the tools-of-the-trade at our disposal.

We were housed in the main building at the Douglas Aircraft Company Santa Monica California plant. Our room was a five meter square vault which previously had been used for storing secret documents. The door was steel with a combination lock. In addition to the mathematical physicist, Leon, we hired an experimentalist Harvey Bjornlie, a retired Los Angeles, California police detective Paul Wilson, a UFO "buff" Stan Friedman, and a psychic Chan P. Thomas.

Chan had been instrumental in promoting the project from its outset. In fact, he is the person who discovered my book *Advanced Physics* and had recommended Bob to hire me. Stan was hired because of his wealth of background on flying saucer sightings. Paul was hired to investigate the flying saucer reports in depth. Bob essentially ran the flying saucer investigative effort, Darell was basically in charge of the experimental efforts, and I was in charge of the theoretical effort. Our budget was less than $100,000.00 per year. The total expenditure of our project in its three year existence was probably less than the US Air Force spent on the project headed by Dr. E.U. Condon [21] trying to prove there were no flying saucers. Incidentally, in the Condon report the researchers found what they believed were flying saucers, but Condon's summary denied their results. In the following paragraphs I will report only the results of the theoretical efforts.

My three year tenure at Douglas was like "dying-and-going-to-heaven". I almost had all my time free to work on the kinetic particle theory of physics. Darell and I searched libraries and all the book stores in the Los Angeles area looking for information useful to the project. I bought and read every book that could be of interest. (I read 1000 books in the three year period.) Leon was an accomplished teacher. He taught me how to use tensors which are required in the analysis of the flow of kinetic particles.

I spent approximately a year developing what I called "the Continuum Equation for Kinetic Particles". In this equation I assumed for each small volume of space considered there were enough particles so that the continuum assumption on space density and velocity density were valid, but did not assume the velocities were Maxwell-Boltzmann distributed. I dropped this effort later when it appeared that the Maxwell-Boltzmann distribution might be valid.

Bob and I went to a seminar on cosmology at the Douglas Research Division. There was some of Arthur Eddington's work discussed on relating cosmological parameters to macroscopic and even microscopic parameters. Bob got the idea he could interrelate all these parameters just using dimensional analysis. He also may have been influenced by the incredible paper by diBertini [22] in which most of the constants of physics were interrelated by a stochastic mathematical structure. The fine structure constant was not mentioned in diBertini's paper nor in the efforts of Bob's. Anyway, Bob and Harvey worked on the dimensional analysis for several months.

I kept cognizant of Bob and Harvey's dimensional analysis and by relating cosmological magnitudes to microscopic magnitudes I related the basic particle (brutino) mass to the mass of the universe and the brutino diameter to the radius of the universe. Anyway, I guessed that the brutino diameter was the Planck length $\left(10^{-35}\,m\right)$ and the mass was $10^{-65}\,kg$, something of the order of the graviton energy divided by the square of the speed of light. This then lead to the idea that the Hubble red shift was due to the loss of one brutino for each wave length which the photon executes during its travel through space.

The bulk of my time the next two years was spent on dreaming up models of the electron, proton, and photon. None of these models withstood the scrutiny of experiment. The only significant milestone reached was the discovery that $\left[\left(\mathrm{v}_r - \mathrm{v}_m\right)/\mathrm{v}_r\right]^2$ was very close to the value of the fine structure constant (one part in 10^3). Since this constant times the speed of light gives the electron orbital velocity which is controlled by inertia, when the speed is adjusted by using a center of mass system the agreement is one part in 10^5 , as discovered by Darell. Bob, Darell, and I all realized this was extremely significant and almost assured us that the kinetic particle theory of physics was valid. We published this result in 1970 [1].

However, hard times came. Douglas Aircraft had been absorbed by McDonnell Aircraft to form McDonnell Douglas Corp in 1967, shortly after my employment. In 1970, the Air Force cancelled the McDonnell Douglas MOL (Manned Orbiting Laboratory) Project. Funds began to be scarce.

McDonnell Douglas had retained Dr. Richard P. Feynman as a consultant and Dr. Feynman was requested to review our project (actually, just the theoretical physics portion). Bob and Darell went to the review. Even in spite of Darell's vehement objections at the meeting, the project was cancelled. Dr. Feynman saw no merit in the work. He ignored the significance of the factor $\left[(\mathrm{v}_r - \mathrm{v}_m)/\mathrm{v}_r\right]^2$ being close to the fine structure constant. Further, he said the center-of-mass correction was not valid. Surely Dr. Feynman knew that this correction was valid. Even further, in his book QED which he published some 15 years later, he discusses the fine structure constant, see Feynman [7], on pages 129 and 130. He referenced Arthur Eddington's numerology giving the value 136 for $1/\alpha$ but he never mentioned our raw value of 137.1087 nor our corrected value of 137.0341 versus the measured value of 137.0360. It appeared that Dr. Feynman felt that we should not be encouraged without first obtaining a proper physics background. In fact, he once told me to "Come up through the ranks."[1] In any case the project was cancelled.

Thus, as a result of three years research we had discovered the numerical connection between the fine structure constant and the kinetic

[1]Many years later I thought that might have been an easier route than what I had taken.

particle ether. (We were not able to discover the mechanism connecting the two until some 30 more years passed.) Also, we knew the mechanism of the Hubble red shift and the mass of the basic particle i.e., the brutino). We had guessed at the particle diameter but had no mechanism for computing its diameter.

I decided to seek a university setting for the continuation of my physics research. I wanted to locate a university in a rural setting where I could teach engineering and work on the kinetic particle theory. I wanted to teach engineering, since I had done that a lot over almost a quarter century prior to the project cancellation. I felt engineering teaching would be intellectually soothing since the kinetic particle research had become stressful.

I located the ideal job as a professor of mechanical engineering at Mississippi State University (MSU). Dr. C.T. Carley employed me to teach mechanical engineering design and to develop research in the department. MSU was a university with 8000 students and it was adjacent to Starkville Mississippi, a town of 16,000 people.

Darell and I immediately published our results to date (1970) in the Journal of the Mississippi Academy of Science [3]. The next break came when I discovered that if particles are taken from a background, aligned without changing their speeds, then squeezed together, if their energy is conserved their linear flow velocity must be increased from the mean speed v_m to the rms speed v_r. This is the mechanism of the neutrino and the speed of the neutrino is $v_r - v_m$ (which is c, the speed of light). However, we still hadn't discovered what the mechanism is that is represented by the ratio $(v_r - v_m)/v_m$ squared.

Our next big break was that Darell became the Director of the Missile Directorate in the US Army Anti-Ballistic Missile Defense Agency in Huntsville, Alabama. I obtained a small research contract from the agency. I employed Leon (Dr. Leon A Steinert) as a consultant. He recognized that the experiments of Bjerknes [6] were applicable to my model of nuclear particles. By then, I had assumed that neutrinos could orbit. This came about as a result of the velocity jump from v_m to v_r producing a large thrust. This thrust was later calculated and its value is more than a million pounds. The orbiting neutrino produced a sinusoidal varying pulse of velocity amplitude $v_r - v_m$ which, at large distances from the particle, acted like the result of a breathing sphere. Since a force in the kinetic particle theory of physics depends upon the square of the velocity, the term $(v_r - v_m)^2$ is proportional to the electromagnetic force. The complete computation of the electromagnetic force F_e is $F_e = \rho_o 2a^4 (v_r - v_m)^2 / R^2$ where ρ_0 is the background density, "a" the particle orbital radius, $v_r - v_m$ is c (the sinusoidal velocity amplitude) and "R" is the separation distance of the two interacting charges.

It was early recognized that the nuclear force was produced by inflows into two nuclear particles at the mean speed of the background gas. However, even now we do not know exactly how the two particles interplay in producing the nuclear force. However, the force is due to the interaction of two orbiting neutrinos (one making up each nuclear particle) and the inflow is sinusoidal with the maximum velocity amplitude v_m (the background mean speed). The nuclear force then is the same as the electrostatic force except with the velocity amplitude being v_m instead of $v_r - v_m$ as in the electrostatic force. This then completes the understanding of the fine structure constant. It is the ratio of the electrostatic force (proportional to $(v_r - v_m)^2$) to the nuclear force (proportional to $v_m{}^2$). Thus the fine structure constant is $\left[(v_r - v_m)/v_m\right]^2$. The understanding of the mechanism for the fine structure constant was completed only in the spring of this year (2004).

Obtaining a complete understanding of the fine structure constant, i.e., the significance of the term $v_r - v_m$, of $(v_r - v_m)^2$, v_m, and $v_m{}^2$ took almost a quarter of a century. It only takes a few minutes to explain it now. The first step in understanding it was the recognition that $v_r - v_m$ was the speed of light. When I discovered the mechanism determining the speed of light (around 1980) I also recognized that the neutrino was an entropy reducing device. Finally I had discovered the basic mechanism which puts organization back into the universe, since all processes previously known to man always result in entropy increases. I published an abstract announcing these two discoveries in 1980, see Brown [23]. This paper was not well received by my colleagues. I thought it would be a simple task to write and solve the dynamic equations proving that the neutrino mechanism resulted from these dynamic equations. However, in the past 25 years I have not been able to solve the equations – nor have I completed an experiment verifying the structure.

The most fundamental place in physics were Planck's constant arises is with the neutrino. The neutrino spins about its axis of translation and thus produces angular momentum. The value of the angular momentum is given by $\hbar = mrv = mr^2\omega \approx 10^{-34}\,kg - m^2/s$, which is Planck's constant. The unusual thing about angular momentum is that it is the same for all neutrinos even though neutrinos have energies which vary more than ten orders of magnitude. How this occurs is that the neutrino has a very small solid core which volume is 10^{-20} times the volume of the neutrino. The angular momentum, within measurement error, is all due to the ether gas flowing into the "neutrino sphere" and turning as it flows in to be able to exit out the one outflow location. This fact was discovered early but the value of the ether gas constants were not known well enough until recently to obtain an accurate calculation of the angular momentum.

Another major puzzle, which had an obvious solution, is what causes the proton to have one precise value of mass. This was solved by placing a single neutrino in orbit and let its large thrusting force balance its centrifugal inertia. At the same time we required that the orbital angular momentum be the same as the neutrino had in its linear path. One, and only one, mass would satisfy these two requirements. A proton thus would be formed by its chance collision with another neutrino having enough mass to knock it into a circular orbit. This was reported in 1985 by Brown [17]. The importance of the proton mass mechanism can not be over emphasized. The proton mass controls the electron mass and the neutron mass. Thus, all the stable matter particle masses are controlled by this one mechanism.

We are not certain how the mass of the electron is determined. Possibly the mass is the amount of excess mass produced when the proton and electron are formed. The electron has a definite mass independent of its field and the only way we have ever conceived of matter at rest is to have the mass exist in an orbital state. The orbital radius is small so the electron structure had to have that path, a path to produce charge, and a third path to produce its angular momentum. Such a structure was conceived and published by Brown [11] in 1991. In fact, the electron structure is displayed on the front cover and dust jacket of this reference.

The structure of the neutron was not discovered until this year (2004) and its first publication is in this book. The neutron is simply the proton and the electron with its angular momentum path removed. The weak nuclear force is simply the result of the chance impact of the proton outflow with its velocity v_r hitting the reduced electron in its neutron path.

Soon after we discovered that the expanding universe was not expanding, that photons were just losing energy during transit, we began to suspect that gravitation was controlled by the action of a single basic ether particle. We eventually determined that the Hubble shift and gravity are produced by the same mechanism. Fundamentally both are caused by the electromagnetic fields of the electron and the proton and are caused by flows that exactly balance each other – except for the size of the basic particle making up the flows. The gravitational model was published on Page 170 of Brown [11] in 1991.

Recently I started re-reading Dr. Feynman's book QED [7] and realized that kinetic particle theory explanations may be readily available for the quantum electrodynamics phenomena described in QED. For example, in the double-slit experiment individual electrons (or photons) pass through a two-hole grating and produce an interference pattern. The explanation is simple, the electron (or photon) field is much larger than the holes and the electron and part of its field goes through one hole and the remainder of the field goes through the other hole. We also found a classical explanation for the reflection of photons from grated mirrors.

During the course of this research to discover the grand unified theory of physics I published 8 books, see references 2, 11, 13-15, and 17-19. The objective of these publications was to garner support for development of the theory. They also helped me to solidify the different concepts throughout the development. Further, I had hoped the publications would reach scientists with interest similar to mine and I would receive comments useful for developing the theory. This hope was hopeless. I received practically no useful feed-back from the publications.

That's the history of the Kinetic Particle Theory of Physics from its inception in 1963 to the present (2004), a 41 year odyssey. Throughout the many years the research effort was mostly part-time. I enjoyed it.

References

1 Brown, J.M., Harmon Jr., D.B., and Wood, R.M., "A Note on the Fine Structure Constant," Mc Donnell Douglas Astronautics Company Paper MDAC WD 1372 Huntington Beach, CA, June 1970.

2 Brown, Joseph M., *Fundamentals of Physics*, ISBN 0-962768-1-0. Basic Research Press, 120 East Main Street, Starkville, MS 39759, 1999.

3 Brown, J.M. and Harmon, Jr. D.B., "A Kinetic Particle Theory of Physics," J. Mississippi Academy of Science, Volume XVIII Pages 1-26, 1972.

4 Mohr, Peter J. and Taylor, Barry N., "The Fundamental Physical Constants" Pages BG6 – BG13, Physics Today Buyers' Guide 2003. August, 2003.

5 Lay, J.E., "An Experimental and Analytical Study of Vortex Flow Temperature Separation by Superposition of Spiral and Axial Flow", Part 1, Pages 202-212, Transactions of the ASME, August, 1959.

6 Whittaker, E.T., *A History of the Theories of Aether and Electricity I*, Pages 284ff. Thomas Nelson and Sons Ltd., London 1951.

7 Feynman, Richard P., *QED The Strange Theory of Light and Matter*, Page 129. ISBN 0-691-08388-6, Princeton University Press, Princeton, NJ 1985.

8 Lévy-Leblond, Jean-Marc, ***Quantics*** *Rudiments of Quantum Physics*, Page 20. North-Holland, NY, 1990.

9 McClusky, S.W., *Introduction to Celestial Mechanics*, Addison-Wesley Publishing Company, Inc., Reading Massachusetts, 1963.

10 Bassett, A.B., *A Treatise on Hydrodynamics*, Vol. 1, pages 248-56, Dover, NY, 1961.

11 Brown, Joseph M., *Principles of Science*, ISBN 0-9626768-0-2 Basic Research Press, 120 East Main Street, Starkville, MS 39759, 1991.

12 Richtmyer, F.K., Kennard, E.H., and Lauritsen, T. *Introduction to Modern Physics*, McGraw-Hill Book Company, Inc., NY, 1955.

13 Brown, Joseph Milroy, *Unified Physics – Part 1*, JMBCo – Research Division, Box 45004, Los Angeles, CA, 90045, 22 November 1965.

14 Brown, Joseph Milroy, *Advanced Physics*, JMBCo – Research Division, Box 45153, Los Angeles, CA, 90045, 18 April 1966.

15 Brown, Joseph Milroy, *Advanced Physics* 3rd Edition,, JMBCo – Research Division, Box 45153, Los Angeles, CA, 90045, 18 April 1967.

16 Binder, R.C., *Fluid Mechanics* 3rd Edition, Prentice-Hall Inc., Englewood Cliffs, NJ, 1955.

17 Brown, Joe, *Brutino Physics*, Sherwood Press, 120 East Main Street, Starkville, MS 39759 USA, 1985.

18 Brown, Joseph M., *Brutino Physics*, 2nd Edition, Sherwood Press, 120 East Main Street, Starkville, MS 39759 USA, 1986.

19 Brown, Joseph M., *Lectures of Physics*, ISBN 0-9626768-2-0, Basic Research Press, 120 East Main Street, Starkville, MS 39759, USA, 1999.

20 Kuhn, Thomas S., *The Structure of Scientific Revolutions*, ISBN 0-226-45803, University of Chicago Press, Chicago, 1996.

21 Condon, E.U., et al., eds. *Scientific Study of Unidentified Flying Objects*, New York, Bantam 1968.

22 diBertini, Robert Oros, "Some Relations Between Physical Constants", P. 737-740, Soviet Physics – Doklady, Vol. 10, No. 8, Feb. 1966.

23 Brown, Joseph M., "A Counterexample to the Second Law of Thermodynamics", abstract, Mississippi Academy of Science, 1980.

24 Frisch, Otto Robert, *What Little I Remember*, ISBN 0-521-28010-9, Cambridge University Press, NY, 1980.

25 Jahnke, Eugene and Emdi, Fritz, *Tables of Function with Formula and Curves*, Page 110, Dover Publications, New York, 1945.

26 Ferrers, N.M., *Spherical Harmonics*, Article 24, London, 1877.

27 Basset, A.B., *A Treatise on Hydrodynamics*, Vol. 2, Pages 2,3. Dover Publications, Inc., New York, 1961.

Index

T

U

V

W